Recent Advances in Chemical Kinetics

Edited by Muhammad Akhyar Farrukh

Published in London, United Kingdom

Recent Advances in Chemical Kinetics
http://dx.doi.org/10.5772/intechopen.98008
Edited by Muhammad Akhyar Farrukh

Contributors
Girish Basavaraju, Y. Ashwin, Prathyusha Naini, Ravishankar Rajanna, Hazem Daoud, Robert E. Coleman, Roger B. Boulton, Alexi A. Stuchebrukhov, Rozina Khattak, Veikko Uahengo

First published in London, United Kingdom, 2022 by IntechOpen
IntechOpen is the global imprint of INTECHOPEN LIMITED, registered in England and Wales, registration number: 11086078, 5 Princes Gate Court, London, SW7 2QJ, United Kingdom

British Library Cataloguing-in-Publication Data
A catalogue record for this book is available from the British Library

Additional hard and PDF copies can be obtained from orders@intechopen.com

Recent Advances in Chemical Kinetics
Edited by Muhammad Akhyar Farrukh
p. cm.
Print ISBN 978-1-80355-225-5
Online ISBN 978-1-80355-226-2
eBook (PDF) ISBN 978-1-80355-227-9

Recent Advances in Chemical Kinetics

Edited by Muhammad Akhyar Farrukh

Published in London, United Kingdom

Recent Advances in Chemical Kinetics
http://dx.doi.org/10.5772/intechopen.98008
Edited by Muhammad Akhyar Farrukh

Contributors
Girish Basavaraju, Y. Ashwin, Prathyusha Naini, Ravishankar Rajanna, Hazem Daoud, Robert E. Coleman, Roger B. Boulton, Alexi A. Stuchebrukhov, Rozina Khattak, Veikko Uahengo

First published in London, United Kingdom, 2022 by IntechOpen
IntechOpen is the global imprint of INTECHOPEN LIMITED, registered in England and Wales, registration number: 11086078, 5 Princes Gate Court, London, SW7 2QJ, United Kingdom

British Library Cataloguing-in-Publication Data
A catalogue record for this book is available from the British Library

Additional hard and PDF copies can be obtained from orders@intechopen.com

Recent Advances in Chemical Kinetics
Edited by Muhammad Akhyar Farrukh
p. cm.
Print ISBN 978-1-80355-225-5
Online ISBN 978-1-80355-226-2
eBook (PDF) ISBN 978-1-80355-227-9

Meet the editor

Dr. Muhammad Akhyar Farrukh is a Professor of Chemistry at the University of Central Punjab, Pakistan. He has more than 20 years of professional and teaching experience in different organizations. He has received many international and national awards, including three gold medals from the University of Karachi, Pakistan, and five gold medals, including one from the Minister of Higher Education Department and the Governor of Punjab. He is also the recipient of the Young Chemist Award from the International Union of Pure and Applied Chemistry (IUPAC), Italy; Young Scientist Awards by The World Academy of Sciences (TWAS), Egypt and InterAcademy Partnership/Global Young Academy (IAP/GYA), Germany; Young Scientist Award from the World Economic Forum, China; Young Researcher Award from the Council for Lindau Nobel Laureate Meetings; and the 2015 IUPAC Award for Chemists as an outstanding chemist in South Korea. Dr. Farrukh established a nano-chemistry lab in Pakistan with grants from national and international organizations.

Contents

Preface

Recent Advances in Chemical Kinetics provides insight into different aspects of chemical reactions. Over six chapters it covers both basic and advanced topics, including solvent catalysis, customized flow reactors, p-conjugation in kinetics, kinetics of autoxidation, molecular chameleon for cations and anions, and internal clocks of atoms.

Chapter 1, "Solvent Catalysis in the Sensitizer-Mediator Redox Kinetics", describes the effect of solvents as an important aspect of the sensitizer-mediator interaction in the reaction medium. It discusses The catalytic role of a small-volume fraction of organic solvent in the aqueous electron transfer kinetics of a few putative sensitizer-mediator reactions is discussed in this chapter. It also investigates the reduction of dicyanobis(2,2'-dipyridyl)iron(III) and dicyanobis(1,10-phenanthroline)iron(III) in binary solvent media, including dilute tertiary butyl alcohol (TBA)-water and dilute 1,4-dioxane-water, the reduction of dicyanobis(2,2'-dipyridyl)iron(III) and dicyanobis(1,10-phenanthroline)iron(III) was investigated.

Chapter 2, "Process Intensification in the Customized Flow Reactors", discusses the design, development, and characterization of continuous flow reactors through RTD studies. It also provides an overview of dimensionless numbers and their influence on the flow reactors, effect of kinetic parameters on batch and flow systems, estimation of various thermodynamic properties to assist equipment design, process intensification of continuous flow reactors to achieve maximum performance, validation of reactors using well-known reaction system and listing out advantages of flow reactors over batch processes.

Chapter 3: "Catalytic Behavior of Extended π-Conjugation in the Kinetics of Sensitizer-Mediator Interaction", discusses the catalytic effect of extended π-conjugation on the electron transfer process between ferricyphen-ferrocyanide and ferricypyr-ferrocyanide in an aqueous medium. Ferricyphen and ferricypyr may be feasible options for the sensitizer in dye-sensitized solar cells due to their high reduction potential, stability, capability as an outer-sphere oxidant, and photosensitivity. This chapter compared the ability of competent putative sensitizers to oxidize the likely mediator in water.

Chapter 4: "The Kinetics of Autoxidation in Wine", explains how tartaric acid can be oxidized in the presence of iron without peroxide in the air. This chapter explores the theoretical considerations of iron complexes formation, oxygen activation, an autoxidative mechanism, and experimental measurements of tartaric acid oxidation as the basis of autoxidation in wine. It discusses the role of iron complexes in the activation of oxygen, the formation of reactive oxygen species, and the initiation of autoxidation, which are crucial for understanding wine oxidation kinetics. Mechanisms based on hydroxyl radicals versus the ferryl species are likely to have different oxidation products of wine components based on pH effects.

Chapter 5: "The ESIPT-Steered Molecular Chameleon for Cations and Anions Based on Alizarin and Alizarin-S: A Comparative Study", deals with the chemosensing properties of Alizarin (A3) and Alizarin S (AS3) toward anions and cations in acetonitrile (CH_3CN). The absorption and fluorescence properties of the two molecular entities were investigated in CH3CN. Based on the excited state intermolecular proton transfer system (ESIPT), the probes were able to collectively discriminate specific cations (Cu^{2+}, Zn^{2+}, Ni^{2+}, Fe^{3+}, Fe^{2+}) and anions (F^-, CN^-, OH^-, AcO^- and N_3^-) via colorimetric observations and spectrometric activities. The simultaneous fluorescence effects were ascribed to the deprotonation activities experienced by A3, as excess anion quantities were added. The sulfonyl electron withdrawing group had an effect in the Alizarin structure, towards the discrimination of anions and cations, both colorimetrically and fluorometrically.

Chapter 6: "Perspective Chapter: Slowing Down the "Internal Clocks" of Atoms – A Novel Way to Increase Time Resolution in Time-Resolved Experiments through Relativistic Time Dilation", discusses how to achieve 2–3 orders of magnitude higher time resolutions than is possible with laser and electron compression technology. The proposed novel method is designed to slow down the "internal clock" of the sample.

I would like to thank the staff at IntechOpen for their cooperation throughout the process of publishing this book.

Dr. Muhammad Akhyar Farrukh
Professor,
Faculty of Science and Technology,
Department of Basic and Applied Chemistry,
University of Central Punjab,
Lahore, Pakistan

Chapter 1

Solvent Catalysis in the Sensitizer-Mediator Redox Kinetics

Rozina Khattak

Abstract

The sensitizer-mediator redox reaction is a vital component of the dye-sensitized solar cells (DSSCs). The efficiency and stability of dye-sensitized solar cells are aided by the kinetics of this redox process. Several reaction parameters influence the kinetics of a reaction, and if those parameters are controlled, the rate of the process and its results can be controlled. One of the most important aspects of the sensitizer-mediator interaction is the reaction medium. Aqueous DSSCs are unquestionably a good replacement when it comes to taking a green approach to avoiding toxic, flammable, and volatile organic solvents and their mixtures, which are commonly used in DSSCs and are known to harm the environment while also reducing the lifetime and stability of the DSSCs. The catalytic role of a small volume fraction of organic solvent in the aqueous electron transfer kinetics of a few putative sensitizer-mediator reactions is discussed in this chapter. In binary solvent media including dilute tertiary butyl alcohol (TBA)-water and dilute 1,4-dioxane-water, the reduction of dicyanobis(2,2′-dipyridyl)iron(III) and dicyanobis(1,10-phenanthroline)iron(III) was investigated. The reactions were carried out in a 10% TBA or dioxane to water media with a volume-volume fraction of both solvents using iodide as a reducing agent. The effect of several parameters on the rate constant was also calculated and analyzed.

Keywords: dye-sensitized solar cells, solvents, kinetics, redox reaction, catalysis

1. Introduction

In the kinetics of reactions, particularly redox reactions, the solvent has a significant effect. Redox reactions occur when two responding entities exchange electrons. The electron giver, or reducing agent, is the one who contributes the electron; the electron acceptor, or oxidizing agent, is the one who accepts the electron. The donation and reception of electrons alter the oxidation states of the reactants since electrons are such small charged particles. As a result, the solvent plays an important role in electron transfer reactions. A few of the most influential characteristics that govern redox reactions include solvation, viscosity, and hydrogen bonding [1]. The solvent organizes and reorganizes itself around the reactants and products before and after the electron transfer event. Similarly, the solvent organizes and reorganizes around the reactants during the production of the transition state [2]. According to the transition state theory of reactions in solution and the double sphere model, the rate constant is related to the dielectric constant of a medium using the following expression [3].

$$lnk = \ln k_0 - \frac{e^2 z_A z_B}{4\pi\varepsilon_0 \varepsilon_r r_\# k_B T} \qquad (1)$$

k and k_0 are the rate constants for any dielectric constant and infinite dielectric constant, respectively, in Eq. (1). The symbols e, z_A, z_B, ε_0, ε_r, $r_\#$, k_B and T represent the constant value of electric charge (a constant in coulombs), charge on reactants A and B, permittivity constant, dielectric constant of medium, inter-nuclear distance between the reacting entities that form the transition state complex, Boltzmann constant, and temperature in Kelvin scale, respectively. The Eq. (1) correlates the rate constant with charges on the reactants and the dielectric constant of the medium in three ways.

1. If the reactants have similar charges, i.e., positive and or negative, the rate constant will drop when the dielectric constant of the medium is decreased.

2. If the reactants have different charges, the rate constant will rise as the dielectric constant of the medium falls.

3. If one or both of the reactants are chargeless or neutral, lowering the dielectric constant of the medium has no effect on the rate constant's value.

Other reaction parameters, such as the effect of ionic strength in a specific reaction, must be zero or close to zero in order to investigate the effect of the dielectric constant on the rate constant and, as a result, the rate of the reaction. Variation in ionic strength has a significant effect on the rate constant of any reaction, and we can find a theoretical value of the rate constant at zero ionic strength by extrapolating the graph to zero ionic strength, i.e., the intercept of the plot [4], using the transition state theory to formulate the primary salt effect. The theoretical value of the rate constant at zero ionic strength is known as the ideal value of the rate constant. The ideal rate constant for a reaction can be calculated in a variety of solvent systems with varied dielectric constants. The dielectric constant of reaction media can be changed by changing the proportion of one solvent to another. To determine the slope of the plot according to Eq. (1), it is advisable to plot the natural logarithm of the ideal rate constant (ln$k$) versus the reciprocal of the dielectric constant ($1/\varepsilon_r$) rather than the natural logarithm of the rate constant at any ionic strength. The value of the inter-nuclear distance between the reactants that constitute the transition state complex and are involved in the rate determining step can be calculated using the slope of the plot. The inter-nuclear distance between the reactants of various reactions can be compared and used to control the kinetics of the reactions under certain experimental conditions.

As a result, it is clear that changing the nature of the reaction media affects the entire electron transfer mechanism. When dealing with the kinetics of redox reactions, it is also worth noting that changing the solvents' proportion in a reaction can change the viscosity and strength of hydrogen bonding, as well as the nature of hydrogen bonding, resulting in either the activation controlled mechanism or the diffusion controlled mechanism [5–12]. It is critical to have precise information on the kinetics of any reaction in order to manage it and make use of it as needed in a process. When it comes to dye-sensitized solar cells (DSSCs), the sensitizer-mediator reaction is crucial to the electron transfer cycle as well as the cell's stability, durability, and efficiency. The influence of variations in the sensitizer on the stability and/or efficiency of DSSCs has been explored by a number of researchers [13–19]. A variety of natural and synthetic dyes were utilized. Others, however, have looked into the effect of mediator variation on DSSC efficiency [20–26]. To avoid flammable, poisonous, and volatile organic solvents, combinations of such solvents with water were tried to increase the DSSCs' stability and efficiency. Various organic solvents and their mixtures have been investigated for this purpose [15, 27, 28]. Aqueous-based dye sensitized solar cells are gaining popularity due to their environmentally beneficial and low-cost characteristics. In order to improve the DSSC's efficiency and stability, aqueous-based sensitizers and electrolytes have recently been explored [29]. The effect of dilute binary solvent media consisting of dilute organic percentage and excess water on two potential photosensitizers to oxidize iodide is described in this chapter. Dicyanobis(2,2′-dipyridyl)iron(III) and dicyanobis(1,10-phenanthroline)iron(III) could be a potential replacement for ruthenium-based dyes due to their stability, solubility, and cost-effectiveness. Furthermore, unlike hazardous ruthenium-based chemicals, the most likely iron-based sensitizers are not environmentally detrimental. The rate constants for the reactions were calculated using 10 volume percent of tertiary butyl alcohol in water and an equal volume percent of 1,4-dioxane in water. When the rate constants were compared, it was discovered that such sensitizer-mediator interactions could have an impact on DSSC's efficiency.

2. Kinetics and solvent catalysis

When the proposed sensitizers, dicyanobis(2,2′-dipyridyl)iron(III) and dicyanobis(1,10-phenanthroline)iron(III), are introduced to the iodide solution in the binary solvent media, the oxidation of iodide is a spontaneous process that does not require any external triggering. When the reduction potential of the redox pair is taken into account, all of the reactions are electrochemically viable [30–32]. The visual color change of the solutions and the spectra of the products show the progress of the reactions and product(s) generation when compared to analogous iron(II) complexes (**Figure 1**) [33, 34]. **Figure 1** shows the wavelength maxima corresponding to dicyanobis(2,2′-dipyridyl)iron(II) and dicyanobis(1,10-phenanthroline)iron(II), which allows time course graphs to be drawn as the absorbance increases as a function of time after the products are formed. The reactions were investigated in the visible region at the wavelength maximum of dicyanobis(2,2′-dipyridyl)iron(II) and dicyanobis(1,10-phenanthroline)iron(II) (**Figure 2**). Each reaction was investigated using a pseudo-order kinetic model with an excess and changing concentration of the mediator, iodide, in comparison to a fixed and low concentration of either dicyanobis(2,2′-dipyridyl)iron(III) or dicyanobis(1,10-phenanthroline)iron(III). The reactions were studied at room temperature in all the reaction media. The oxidant concentration was held constant at 0.08 mM and the iodide (reductant) concentration was adjusted between 0.08 mM and 4 mM at 1:1, 1:2.5, 1:5, 1:7.5, 1:10, 1:20, 1:30, 1:40, and 1:50 times to preserve the pseudo-first order kinetic model in all reaction media at 0.06 M ionic strength (μ). The integration method was implemented on the absorbance data, and zero order kinetics was observed corresponding to the sensitizers i.e., oxidizing agents. The absorbance was plotted against time that yielded a slope "$\varepsilon \cdot b \cdot k_{obs}$" that is the multiplication product of the molar absorptivity of either of dicyanobis(2,2′-dipyridyl)iron(II) and or dicyanobis(1,10-phenanthroline)iron(II),

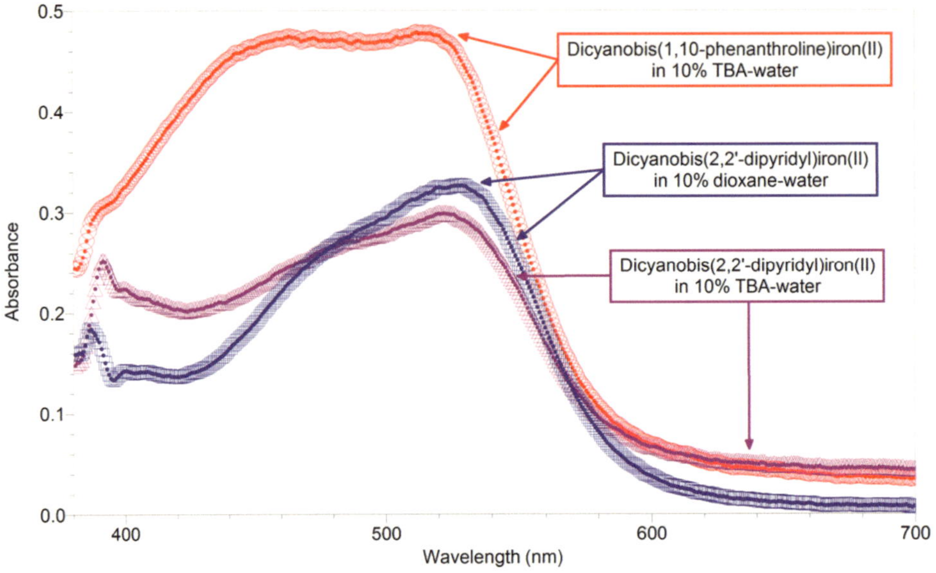

Figure 1.
Visible absorption spectra of the products of the redox reactions in three different reaction media.

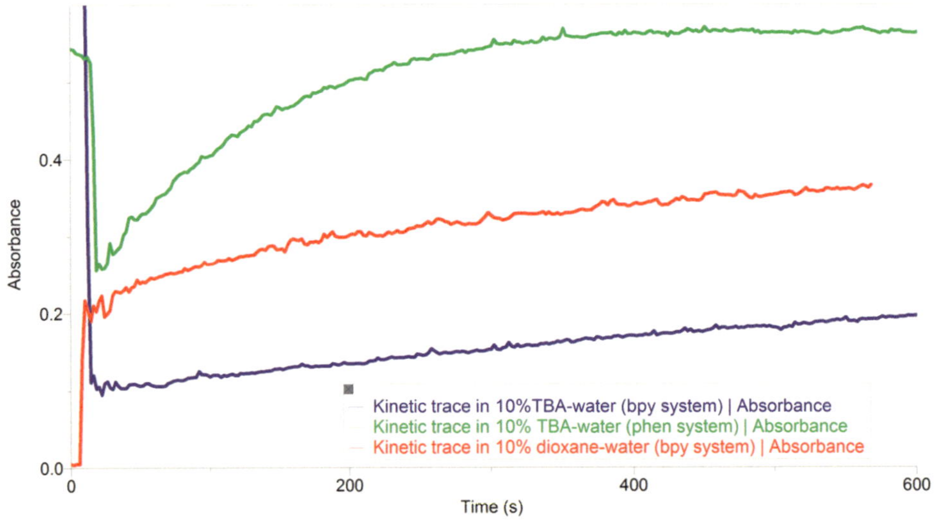

Figure 2.
Representative time course graphs of reactions in different reaction media at 293 ± 1 K.

the pathlength of the quartz cuvette (1 cm) and the observed zero order rate constant, respectively. The slope is a constant (ε•b) times larger than the real value due to the inclusion of a constant mathematical number, which has no overall effect on the rate constant. Such a slope can be obtained by using absorbance without converting it to concentration (via implementing Beer-Lambert's law). When a graph is drawn between the concentrations as a function of time, it yields a straight line with a slope equal to the zero order rate constant. **Figure 2** shows representative kinetic traces at a 1:5 sensitizer:mediator ratio for comparative examination in various solvent media. In 10% (v/v) TBA-water, the reaction between dicyanobis(2,2′-dipyridyl)iron(III) and iodide is the slowest, but dicyanobis(1,10-phenanthroline)iron(III)-iodide is the fastest in identical media.

Controlling the reaction within the DSSC with respect to potential sensitizers such as dicyanobis(2,2′-dipyridyl)iron(III) or dicyanobis(1,10-phenanthroline)iron(III) is aided by zero order kinetics corresponding to oxidizing agents in all reaction media. The mediator, such as iodide, plays the main role in controlling the reaction kinetics in such sensitizer-mediator interactions. When the reaction mechanism is known in all of the selected media, it becomes much easier to exploit this sensitizer-mediator interaction to get the most out of the reaction in a DSSC where the rate of the reaction is solely dependent on the mediator. The zero order rate constant obtained for each reaction in all reaction media was displayed as a function of the iodide ion concentration (**Figure 3**) for this experiment [4, 35, 36]. The redox reaction between dicyanobis(2,2′-dipyridyl)iron(III)-iodide in either 10% TBA-water (bpy-TBA in **Figure 3**) or 10% dioxane-water (bpy-dioxane in **Figure 3**) underwent a first order with the zero order rate constant increasing linearly with increasing iodide concentration, yielding a straight line passing through the origin. The overall first-order rate constant of the reaction is determined by the slope of the plot. In the meantime, a third order kinetics was found in the reaction of dicyanobis(1,10-phenanthroline)iron(III)-iodide in 10% TBA-water (phen-TBA in **Figure 3**). As a result, it has been discovered that in the selected reaction media, dicyanobis(1,10-phenanthroline)iron(III)-iodide reacts much faster than dicyanobis(2,2′-dipyridyl)iron(III)-iodide, implying that in a DSSC, the recombination process may be faster in the "phen" system rather than the "bpy" system. To avoid repeating long names, the former potential sensitizer is referred to as "bpy" and the latter as "phen". The main difference

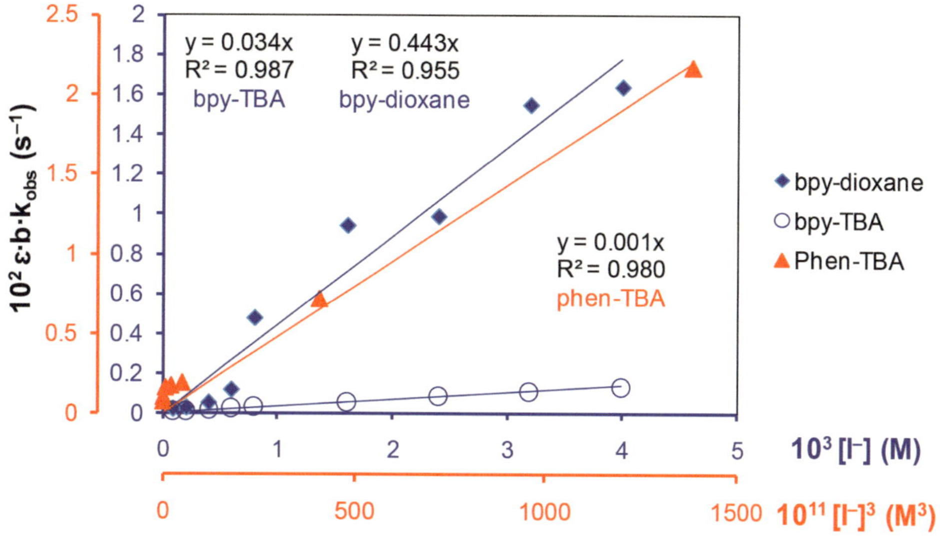

Figure 3.
Kinetic study with respect to the reducing agent in different reaction media.

between the two sensitizers is in the chelate, where the phen system has more pi-conjugation than the bpy system. The rest of the coordination sites and geometry, on the other hand, are similar. Both are octahedral complexes that are substitution inert. Furthermore, in the instance of dicyanobis(2,2′-dipyridyl)iron(III)-iodide, the first order rate constant is 13 times bigger in 10% dioxane-water than in 10% TBA-water. Consequently, the results reveal that dioxane has a catalytic influence on the redox kinetics of the sensitizer-mediator relationship when compared to TBA. Similarly, dicyanobis(1,10-phenanthroline)iron(III)-iodide displays faster electron transfer than dicyanobis(2,2′-dipyridyl)iron(III)-iodide in the identical reaction medium (10% TBA-water).

To evaluate the catalytic function of the solvent in the sensitizer-mediator interaction, it is obvious to calculate the ideal rate constant for each reaction in each solvent medium, such as the rate constant at zero ionic strength. **Figure 4** depicts the plots of the primary salt effect on the rate constant according to the formulation (2). The optimal value of the rate constant was determined by the intercept of the plots. When the effect of the solvent is significant and there is no effect of ions on the rate constant, the ionic strength is assumed to be zero, and the ideal value of the rate constant is obtained. When the experimental data is plotted and extrapolated to zero ionic strength, the theoretical value of the rate constant, or ideal rate constant, is produced.

$$\log(\varepsilon \cdot b \cdot k_{obs}) = \log(\varepsilon \cdot b \cdot k_{obs})_{ideal} + 2Az_A z_B \frac{\sqrt{\mu}}{1+\sqrt{\mu}} \tag{2}$$

Figure 4 shows the decelerating effect of increasing ionic strength on the observed zero order rate constant, indicating that opposite charges are involved in

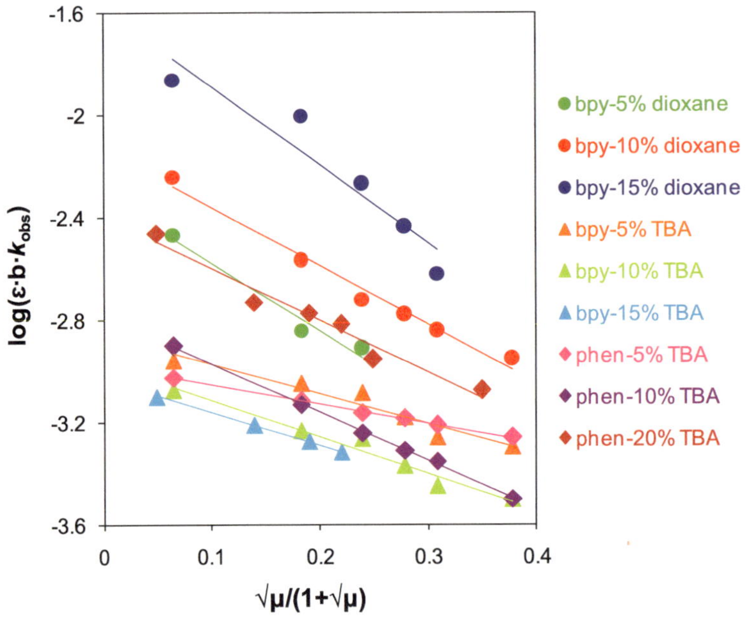

Figure 4.
Plots of primary salt effect in different solvent media.

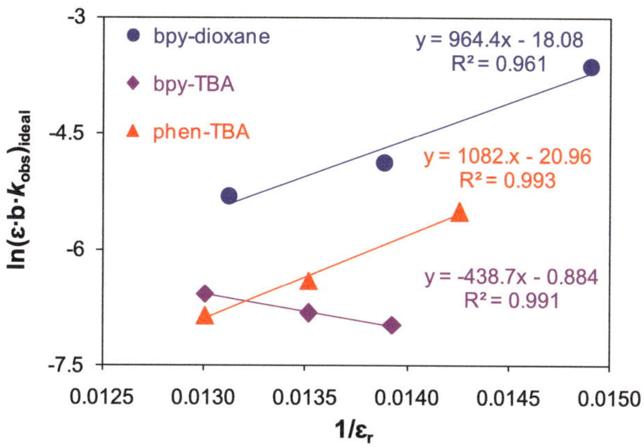

Figure 5.
Effect of solvent on the reaction kinetics of sensitizer-mediator interaction.

the rate determining step that leads to the formation of the transition state complex. bpy-5%/10% /15% dioxane, bpy-5%/10%/15% TBA, and phen-5%/10%/20% TBA were used to depict the effect of increasing ionic strength in different solvent media for "bpy" and "phen" systems. The term bpy-5% dioxane, on the other hand, refers to a sensitizer-mediator interaction involving dicyanobis(2,2′-dipyridyl) iron(III)-iodide and a 5% (v/v) 1,4-dioxane-water solvent system. Meanwhile, dicyanobis(1,10-phenanthroline)iron(III)-iodide in a 5% (v/v) TBA-water solvent solution is phen-5% TBA. The remainder of the terms has comparable connotations as well. The ideal value of the rate constant was obtained from the intercept of each plot and was used to build a graph. According to Eq. (1), the natural logarithm of the ideal rate constant was drawn on the y-axis and the reciprocal of the dielectric constant was drawn on the x-axis for each system, including bpy-dioxane, bpy-TBA, and phen-TBA. **Figure 5** depicts the final results. The slope of the plots was used to calculate the inter-nuclear distance ($r_{\#}$) between the active species that constitute the transition state complex, and the results are presented in **Table 1**. **Table 1** demonstrates that in the reaction of dicyanobis(2,2′-dipyridyl)iron(III) with iodide in 10% (v/v) TBA-water versus 10% (v/v) dioxane-water, the inter-nuclear distance is very long enough. This exhibits the catalytic impact of dioxane over TBA by displaying the quick electron transfer kinetics between the sensitizer-mediator in dioxane-water as compared to TBA-water. The inter-nuclear distance between the active reactants that form the transition state complex and lead to the rate-determining step of the reaction in dicyanobis(1,10-phenanthroline)iron(III)-iodide is smaller (53 pm) than dicyanobis(2,2′-dipyridyl)iron(III)-iodide, which is 130 pm in 10% (v/v) TBA-water. This confirms that in the former situation, electron transfer between the sensitizer and mediator is faster than in the latter case, and that the solvent has a catalytic impact in the "phen" system rather than the "bpy" system. However, by utilizing 1,4-dioxane instead of tertiary butyl alcohol, the reaction of the "bpy" system accelerated almost to the level of the "phen" system, where $r_{\#}$ is 59 pm in the former instance and 53 pm in the latter case. As a result, in order to accelerate a sensitizer-mediator interaction for rapid recombination in DSSC, solvent can be used in an environmentally friendly and cost-effective manner to increase the stability and efficiency of the solar cell.

Sensitizer-Mediator	Solvent System (Reaction Media)	Inter-nuclear Distance ($r_{\#}$) in pm
$[Fe(bpy)_2(CN)_2]^+$- I^-	10% (v/v) TBA-water	130
$[Fe(bpy)_2(CN)_2]^+$- I^-	10% (v/v) dioxane-water	59
$[Fe(phen)_2(CN)_2]^+$- I^-	10% (v/v) TBA-water	53

Table 1.
Catalytic effect of solvent in the sensitizer-mediator interaction.

3. Conclusion

The redox reaction of dicyanobis(2,2′-dipyridyl)iron(III)-iodide and dicyanobis(1,10-phenanthroline)iron(III)-iodide could be used to improve the stability and efficiency of dye-sensitized solar cells. The proposed sensitizer-mediator interaction could be both cost-effective and ecologically advantageous due to the low cost of photosensitive iron complexes. To catalyze the indicated sensitizer-mediator processes in aqueous medium, a modest amount (10 volume percent) of organic solvent, such as tertiary butyl alcohol and 1,4-dioxane, can be utilized. Both solvents are relatively benign to the environment and inert to involvement in the redox process, so they do not cause parallel reactions or complex kinetics. In contrast to dicyanobis(2,2′-dipyridyl)iron(III)-iodide, which follows first order, tertiary butyl alcohol catalyzes the reaction of dicyanobis(1,10-phenanthroline)iron(III)-iodide to third order. Both reactions have rates that are independent of the concentration of oxidizing agents (potential sensitizers), emphasizing the need to use such oxidants in DSSCs to control a reaction that is solely dependent on the concentration of the reductant/reducing agent (mediator). Such sensitizer-mediator interactions are simple to manage, and the rate of reaction can be sped up or slowed down by adjusting the concentration of just one reactant. Furthermore, the rate of reaction was enhanced several times faster by using 1,4-dioxane instead of tertiary butyl alcohol, which may be useful in the recombination process in DSSCs. The inter-nuclear distances for the bpy-TBA and phen-TBA systems were 130 and 53 pm, respectively. However, with the bpy-dioxane and phen-TBA systems, it was 59 and 53 pm, respectively. These findings pertain to solvent catalysis by reducing inter-nuclear distances, resulting in a rapid electron transfer process. This research/study contributes to the development of a cost-effective and environmentally friendly strategy for improving the stability and efficiency of DSSCs.

Author details

Rozina Khattak
Department of Chemistry, Shaheed Benazir Bhutto Women University, Peshawar, Pakistan

*Address all correspondence to: rznkhattak@yahoo.com; rznkhattak@sbbwu.edu.pk

References

[1] Matamoros-Fontenla MS, López-Cornejo P, Pérez P, Prado-Gotor R, Rdl V, Moyá ML, et al. A study of the electron-transfer reaction between $Fe(CN)_2(bpy)_2$ and $S_2O_8^{2-}$ in solvent mixtures: The translational component of solvent reorganization. New Journal of Chemistry. 1998;**22**:39-44. DOI: 10.1039/A706747J

[2] Marcus RA. Electron transfer reactions in chemistry. Theory and experiment. Pure & Applied Chemistry. 1997;**69**(1):13-29

[3] Wright MR. An Introduction to Chemical Kinetics. England: John Wiley & Sons Ltd; 2004

[4] Khattak R, Khan MS, Ullah R, Zainab S, Ali M, Rahman W, et al. Effect of the ionic strength on the redox reaction of dicyanobis(bipyridine) iron(III)-iodide in binary and ternary solvent systems. International Journal of Chemical Kinetics. 2021;**53**(1):16-26

[5] Zhao Y, Freeman GR. Solvent effects on the reactivity of solvated electrons with ions in tert-butanol/water mixtures. Canadian Journal of Chemistry. 1995;**73**(3):392-400

[6] Walker TW, Chew AK, Li H, Demir B, Zhang ZC, Huber GW, et al. Universal kinetic solvent effects in acid-catalyzed reactions of biomass-derived oxygenates. Energy & Environmental Science. 2018;**11**(3):617-628

[7] Besbes R, Ouerfelli N, Latrous H. Density, dynamic viscosity, and derived properties of binary mixtures of 1,4 dioxane with water at T=298.15 K. Journal of Molecular Liquids. 2009;**145**(1):1-4

[8] Gonzalez AG, Pablos F. Evaluation of acidity constants in dioxane-water mixtures by spectrophotometric and potentiometric pH titrations. Analytica Chimica Acta. 1991;**251**:321-325

[9] Ben-Naim A, Yaacobi M. Hydrophobic interaction in water-p dioxane mixtures. The Journal of Physical Chemistry. 1975;**79**(13):1263-1267

[10] Ying Guang W, Masaaki T, Toshiyuki T. A local solvent structure study on 1,4-dioxane-water binary mixtures by total isotropic Rayleigh light scattering method. Journal of Molecular Liquids. 2001;**94**(3):273-282

[11] Cortijo M, Llor J, Sanchez-Ruiz JM. Thermodynamic constants for tautomerism, hydration, and ionization of vitamin B_6 compounds in water/dioxane. The Journal of Biological Chemistry. 1988;**263**(34):17960-17969

[12] Asensio SB, Lopez-Cantarero E, Llor J. The thermodynamics of the ionization of 3-hydroxypyridine and pyridoxine in water-dioxane mixtures. Canadian Journal of Chemistry. 1992;**70**:1635-1639

[13] Mukherjee S, Bowman DN, Jakubikova E. Cyclometalated Fe(II) complexes as sensitizers in dye-sensitized solar cells. Inorganic Chemistry. 2015;**54**(2):560-569

[14] Venkatraman V, Raju R, Oikonomopoulos SP, Alsberg BK. The dye-sensitized solar cell database. Journal of Cheminformatics. 2018;**10**(1):18

[15] Ramirez-Perez J, Maria C, Santacruz CP. Impact of solvents on the extraction and purification of vegetable dyes onto the efficiency for dye-sensitized solar cells. Renewables: Wind, Water, and Solar. 2019;**6**(1):1

[16] Ferrere S. New photosensitizers based upon $[Fe(L)_2(CN)_2]$ and $[Fe(L)_3]$ (L) substituted 2,2′-Bipyridine: Yields for the photosensitization of TiO_2 and effects on the band selectivity. Chemistry of Materials. 2000;**12**:1083-1089

[17] Duchanois T, Liu L, Pastore M, Monari A, Cebrián C, Trolez Y, et al. NHC-based iron sensitizers for DSSCs. Inorganics. 2018;**6**(2):63. PubMed PMID. DOI: 10.3390/inorganics6020063

[18] Li C-T, Lin RY-Y, Lin JT. Sensitizers for aqueous-based solar cells. Chemistry – An Asian Journal. 2017;**12**(5):486-496

[19] Rodrigues RR, Peddapuram A, Dorris AL, Hammer NI, Delcamp JH. Thienopyrroledione-based photosensitizers as strong Photoinduced oxidants: Oxidation of Fe(bpy)32+ in a >1.3 V dye-sensitized solar cell. ACS Applied Energy Materials. 2019;**2**(8):5547-5556

[20] Boschloo G, Hagfeldt A. Characteristics of the iodide/triiodide redox mediator in dye-sensitized solar cells. Accounts of Chemical Research. 2009;**42**(11):1819-1826

[21] Yang K, Yang X, Zhang L, An J, Wang H, Deng Z. Copper redox mediators with alkoxy groups suppressing recombination for dye-sensitized solar cells. Electrochimica Acta. 2021;**368**:137564

[22] Boschloo G. Improving the performance of dye-sensitized solar cells. Frontiers in Chemistry. 2019;**7**(77):1-9. DOI: 10.3389/fchem.2019.00077

[23] Saygili Y, Stojanovic M, Flores-Díaz N, Zakeeruddin SM, Vlachopoulos N, Grätzel M, et al. Metal coordination complexes as redox mediators in regenerative dye-sensitized solar cells. Inorganics. 2019;**7**(3):30. PubMed PMID. DOI: 10.3390/inorganics7030030

[24] Torben Jost D. Novel Redox Couples for Dye-Sensitized Solar Cells Monash University. Thesis. 2017. DOI: 10.4225/03/5898143e7401e. Available from: https://bridges.monash.edu/articles/thesis/Novel_redox_couples_for_dye-sensitized_solar_cells/4621912

[25] Bignozzi CA, Argazzi R, Boaretto R, Busatto E, Carli S, Ronconi F, et al. The role of transition metal complexes in dye sensitized solar devices. Coordination Chemistry Reviews. 2013;**257**(9):1472-1492

[26] Dissanayake MAKL, Umair K, Senadeera GKR, Kumari JMKW. Effect of electrolyte conductivity, co-additives and mixed cation iodide salts on efficiency enhancement in dye sensitized solar cells with acetonitrile-free electrolyte. Journal of Photochemistry and Photobiology A: Chemistry. 2021;**415**:113308

[27] Hsu H-L, Hsu W-T, Leu J. Effects of environmentally benign solvents in the agarose gel electrolytes on dye-sensitized solar cells. Electrochimica Acta. 2011;**56**(17):5904-5909

[28] Kafafy H, Wu H, Peng M, Hu H, Yan K, El-Shishtawy RM, et al. Steric and solvent effect in dye-sensitized solar cells utilizing phenothiazine-based dyes. International Journal of Photoenergy. 2014;**2014**:548914

[29] Kim J-H, Kim D-H, So J-H, Koo H-J. Toward eco-friendly dye-sensitized solar cells (DSSCs): Natural dyes and aqueous electrolytes. Energies. 2022;**15**(1): 219. PubMed PMID. DOI: 10.3390/en15010219

[30] Khattak R, Naqvi II, Summer S, Sayed M. Mechanism of the oxidation of 1-(ferrocenyl)-ethanone/ethanol by dicyanobis(phenanthroline) iron(III). Arabian Journal of Chemistry. 2019;**12**(8):4240-4250

[31] Wang X, Stanbury DM. Copper catalysis of the oxidation of iodide by $[Fe^{III}(bpy)_2(CN)_2]^+$ in acetonitrile. The Journal of Physical Chemistry. A. 2004;**108**(38):7637-7638

[32] Wang X, Stanbury DM. Direct oxidation of l-cysteine by $[Fe^{III}(bpy)_2(CN)_2]^+$ and $[Fe^{III}(bpy)(CN)_4]^-$. Inorganic Chemistry. 2008;**47**(3):1224-1236

[33] Khattak R. Comparative Kinetic Study for the Electron Transfer Reactions of some Iron Complexes [PhD]. Karachi: University of Karachi; 2011

[34] Schilt AA. Mixed ligand complexes of iron(II) and (III) with cyanide and aromatic di-imines. Journal of the American Chemical Society. 1960;**82**(12):3000-3005

[35] Khattak R, Khan MS, Iqbal Z, Ullah R, Khan A, Summer S, et al. Catalytic effect of 1,4-Dioxane on the kinetics of the oxidation of iodide by Dicyanobis(bipyridine)iron(III) in water. Catalysts. 2021;**11**(7):840. DOI: 10.3390/catal11070840

[36] Khattak R, Khan MS, Summer S, Ullah R, Afridi H, Rehman Z, et al. Kinetics of the oxidation of iodide by dicyanobis(phenanthroline)iron(III) in a binary solvent system. International Journal of Chemical Kinetics. 2021;**53**(2): 230-241

Chapter 2

Process Intensification in the Customized Flow Reactors

Girish Basavaraju, Y. Ashwin, Prathyusha Naini and Ravishankar Rajanna

Abstract

The development and intensification of flow reactors are emerging as a new niche area in the pharmaceutical process development. However, development without adequate knowledge of chemistry, kinetics, mass and heat transfer, equipment design, residence time and residence time distribution (RTD) aspects could be difficult to realize the true potential of flow processing development. Understanding these aspects would enable the experimenter to explore options, troubleshoot, and find insights into a wide array of possibilities for continuous flow chemistry development and scaleup. The study provides a systematic approach for flow process development, insights to characterize the equipment, optimize the conditions, and derive meaningful conclusions. The results are presented through appropriate tables and charts to understand the concepts and apply them for various processes. As a case study, a well-known saponification reaction was subjected to represent the performance of the batch versus the flow process (packed bed reactor and tubular reactor) and listed advantages.

Keywords: process intensification, packed bed reactor, tubular reactor, kinetics, saponification of ethyl acetate

1. Introduction

1.1 Highlights

- Designing, development, and characterization of continuous flow reactors (tubular and packed bed reactor) through RTD studies.
- Overview of dimensionless numbers and their influence on these flow reactors
- Effect of kinetic parameters over batch and flow systems
- Estimation of various thermodynamic properties to assist equipment design
- Process intensification of continuous flow reactors to achieve maximum performance.

- Validation of these reactors using well know reaction system and listing out advantages of flow over batch processes.

Continuous production techniques have been used by the chemical industry for a long time, but it is only recently that flow equipment has become available for the use of the laboratory scale, especially in the pharmaceutical industry. This means that flow processes established in the lab could be readily transferred to the production facilities and scaled for commercial use, without substantially altering the reaction conditions [1–3]. The flow processes are gaining high visibility across pharmaceutical industries for varied reasons [4]. One reason could be predominately the economics of running the batch processes verses the flow process. At the onset, the flow processes are well established in the manufacturing of commodity chemicals, meanwhile, the batch processes are highly acquainted in the pharmaceutical industries. In the advent, the flow processes offer wide advantages such as effective heat and mass transfer, superior inherent safety, flexibility, reproducibility, energy efficiency, high reactor throughput, fast and effective mixing, low footprint, in-line automation, and low operating cost [5].

Flow processing has demonstrated chemical production safer, more reproducible, and scalable while offering reduced cost and low environmental impact. Flow processes are more energy-efficient, with precise control over reaction conditions leading to less waste and environmental impact and serving green chemistry principles [6].

By way of example, for every kilogram of a fine chemical produced by the pharmaceutical industry, 5–100 times that amount of chemical waste is being generated [7]. This unacceptable inefficiency with the present state-of-the-art, large scale batch production of chemicals is driving the adoption of resource-efficient flow chemistry alternatives as innovative solutions for chemical manufacturing.

Developments are at a high pace in transforming the batch chemistries to flow processes at the academic level and there is a quite demand building up across industries. In recent years, flow chemistry has become a viable alternative to traditional batch chemistry, with a six-fold increase [6] in the publications featuring micro and meso reactors. The literature which supports the transformations was more in running the experiments without the engineering concepts being discussed such as kinetics, mixing, dispersion, and residence time distributions. On the other side, there have been numerous companies launching flow process development skids for quick and easy development strategies without insight on the reaction or its suitability. An effort towards understanding these concepts become decisive.

Flow reactors for continuous flow processing are typically tubular, packed bed, or microfluidic chip-based systems, where reagents are introduced at different points into the tube in a continuous stream [8]. Because of the small dimension of the tubes and built-in automation, well-defined temperature, pressure, and reaction times are achieved thereby achieving desired product profiles. Initial capital outlay is reduced, compared to traditional batch reactors, and scale-ups could be achieved by running identical parallel channels, making flow chemistry a viable manufacturing approach for small and niche manufacturers.

The characterization of the reactor such as flow patterns becomes essential to decide the performance of these reactor types. There are specific methods available in the literature to characterize the flow reactors, whether it is plug flow, CSTR (Continuous stirred tank reactors), fluidized bed reactor, or packed bed reactor [9–11]. Generally, there is two class of reactors, which are completely mixed or completely plug flow reactors. The residence time distribution (RTD) studies [12–15] were performed to characterize [16–19] the reactor types and to estimate the deviation from the ideal behavior of the reactor under the flow conditions. All the real reactors fall, somewhere between mixed and plug flow [12, 20], the reason could be due to stagnation, recycling of the fluid, channeling of fluid, the difference in the temperature, inadequate mixing with the reactant streams and axial dispersion patterns [13].

As a first approximation, one could establish a model around each reactor to prove the performance using the characteristic information defined in the literature [21]. In the real scenario, we could realize the ineffective contacting, mixing, and lowering in the performance than the ideal case [18, 22]. The RTD is characteristic information to estimate the degree of mixing and opportunities to improve the same through the design of the reactor [12]. Nevertheless, the RTD studies provide significant information around the gaps and opportunities to improve the process from an equipment perspective [22–24]. In general, the shortfalls could be around the channeling, recycling of the fluid, stagnation, and dead zones within the reactor [9].

Various dimensionless parameters were discussed in the literature to support the studies and to develop correlation to understand the performance behavior of these reactors [22, 25–27].

In the present study, customized plug flow and packed bed reactors were designed and fabricated on an appropriate scale. The reactors were characterized through detailed RTD studies. The characteristic plots are used to estimate the behavior of the flow reactor. Well-studied saponification of ethyl acetate in the presence of sodium hydroxide was considered for validating these reactors and demonstrating the advantages of flow processes over the batch process. The hydrolysis of ethyl acetate was essentially an irreversible second-order reaction, in which the sodium acetate and ethyl alcohol were formed as products. In the literature, the emphasis was given to reaction kinetics and mechanism of the reaction than process intensification using various reactor types and their importance. The information around detailed process intensification studies is very nominal and not available in the open literature to the best of authors' knowledge.

2. Materials and methods

All the chemicals were sourced from Sigma Aldrich, Bengaluru, and used for the complete experimental work. The stock solutions were prepared freshly every time using Milli-Q water before conducting any experiments, to avoid data inconsistency.

Conductivity meter (Eutech, CyberScan CON 510, supplied by ComBiotech, Bengaluru) was used to analyze the concentration of sodium hydroxide solution. The customized reactors were built using materials sourced from ComBiotech-Bengaluru, Huber-Germany, and Swagelok-India respectively. We have investigated the saponification reaction as a general procedure for the process intensification in these customized flow reactors.

3. Experimental section

3.1 Calibration curve for the concentration analysis

Calibration curve was established using freshly prepared stock solution of sodium hydroxide of various concentrations such as 40, 80, 400, 800, 2000–4000 mg/L. The conductivity of each stock solution was measured using a conductivity meter. The values of conductance of each sample were plotted against the concentration of sodium hydroxide, which was used as a reference for all the experiments to find the unknown concentration of experimental samples.

3.2 Equipment setup and experimental procedure

3.2.1 Batch reactor setup

A schematic representation of the batch reactor setup is shown in **Figure 1**. All the batch experiments were performed in a jacketed reactor setup equipped with a 250 mL reactor fitted with a coiled condenser. The pitch blade turbine was used as an agitator for conducting experiments. The head of the reactor had 5 ports, of which one was used for a temperature probe (T), other for the agitator shaft, a third for condenser along with nitrogen vent, fourth was used for inserting conductivity probe (C), and the fifth one for dosing reagents.

3.2.2 Tubular reactor setup (type 1)

The tubular flow reactor was fabricated using perfluroalcoxy alkane (PFA) tubes with internal diameter 1.58 and 3.175 mm (supplied by Swagelok) coiled with the definite volume rolled into a disc form, connected with a T- joint and a static mixer (supplied by ColeParmer) and immersed into a bath circulator. Two pumps A and B (QG50, supplied by FMI, Inc) were used to pump the reaction mixture at a definite flow rate. The open bath circulator (supplied by Huber, Germany) was used to control the temperature of the flow reactor. The schematic and actual image of the experimental setup is shown in **Figure 2**.

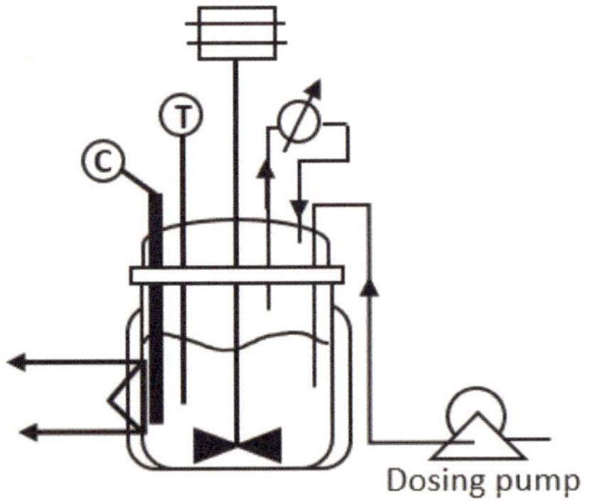

Figure 1.
Schematic and actual image of the batch reactor.

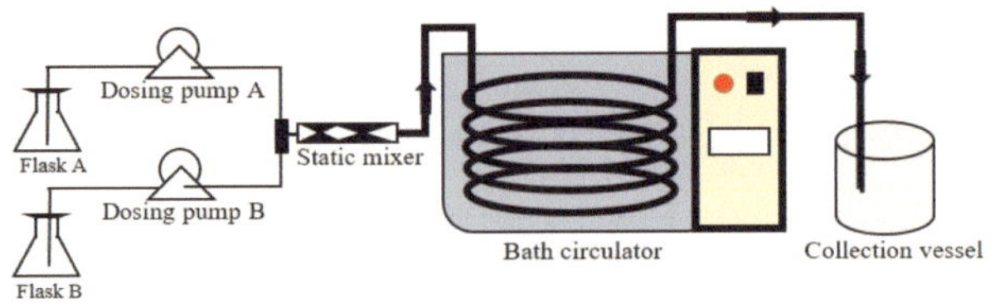

Figure 2.
Schematic and actual image of the tubular reactor.

3.2.3 Packed bed reactor (type 2)

The packed bed reactor (fabricated through ComBiotech, Bengaluru) was made up of glass, filled with glass beads of 2 mm thickness with a sintered disc at the bottom. The diameter and the length of the reactor are 24 and 240 mm respectively. The volume of the column is 105 mL, with packing is around 56 mL. The PFA tubes were used for all fluid connections. The reactor jacket inlet and outlet (12 mm threaded ends) were connected to a bath circulator (supplied by Huber, Germany), to maintain the required temperature of the reactor. The reactor inlet was connected to two pumps (QG50, Fluid metering, Inc) to pump the reaction mixture along with a static mixer in-line to the reactor and a k-type thermocouple was placed in the packed bed to monitor the internal bed temperature. The schematic and actual image of the experimental setup is shown in **Figure 3**. The output of the reactor was connected to a collection vessel. The samples were drawn from the collection tube and analyzed through a conductivity meter. The reactions were performed both in an upward and downward flow to study the process variations in detail.

3.2.4 Tubular reactor setup (type 3)

The tubular reactor was fabricated using PFA tubes with a diameter 1.58 mm (supplied by Swagelok) coiled with the definite volume rolled into a disc form, connected with a T- joint and a static mixer (supplied by ColeParmer) and immersed into a temperature-controlled Sonicator bath. Two pumps A and B were used to pump the reaction mixture at a definite flow rate. The Sonicator bath (supplied by ComBiotech, Bengaluru) was used to control the temperature of the flow reactor with a fixed sonication frequency. The output of the reactor was connected to a collection vessel.

3.2.5 RTD studies to characterize the reactor types

The flow reactors were characterized through RTD studies using sodium hydroxide (NaOH) as tracer input to develop characteristic curves such as E, F, and C curves and other relevant parameters. For better visual detection and quantification of flow patterns of the tracer element, Rhodamine B (synthetic dye) was also used. The concentration of the tracer was around 0.1 N NaOH, prepared through dissolving 4 grams of NaOH in 1 liter of Milli-Q water.

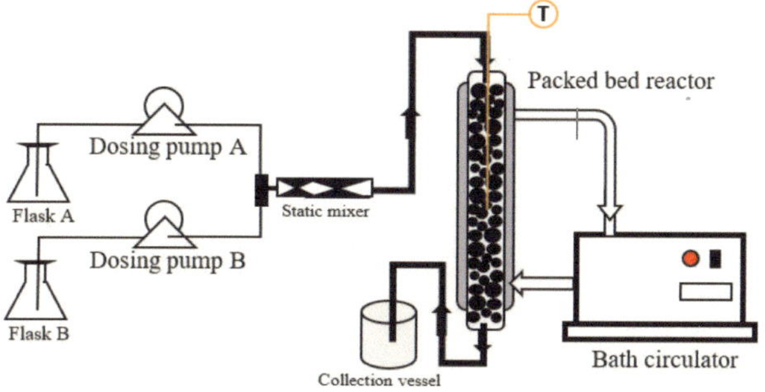

Figure 3.
Schematic and actual image of the packed bed reactor.

Two methods of injection were used, such as pulse and step inputs during the experiments. The pulse input injection was done by the construction of a T-joint and an injection port, where a 10 mL tracer was injected. For the step input, the inlet of the pump was swapped from inline water flow into the beaker containing the tracer solution at time t = 0.

A constant flow rate of 15.8 mL/min was maintained for all the trials performed. The tracer concentration at the exit was measured by a conductivity meter as a function of time and plotted all the characteristic curves. The concentration of tracer measured was in terms of milli siemens per centimeter.

3.3 Estimation of reaction kinetics and relevant parameters

3.4 Reaction scheme

The saponification reaction under basic condition is represented by the following reaction scheme. The rate equation for this reaction could be represented as (**Figure 4**) [28].

$$-r_a = -r_{NaOH} = k\, C_{NaOH}\, C_{Ethylacetate}.$$

The hydrolysis of ethyl acetate was essentially an irreversible second-order reaction, in which the sodium acetate and ethyl alcohol were formed as products. The reaction was represented as ethyl acetate proceeds with a direct attack of the nucleophile on the carbon atom of ethyl acetate [25]. In another study [16], found transition complex formation was a result of nucleophilic interaction of water molecule, where hydroxide ions generally assist the complex formation. The reaction investigation was conducted through a series of experiments in a tubular, packed, and batch reactor and analyzed. The experimental data were fitted with a second-order model at the end to compare the performance [17].

4. Results and discussion

A calibration curve was established using a freshly prepared stock solution of sodium hydroxide of various concentrations. For reaction samples, at time zero, the conductivity values were measured before the reaction was initiated to get the initial concentration of the solutions and then measured at each time interval in batch. Similarly, the samples from the exit of the flow reactors are measured to obtain the amount of sodium hydroxide reacted with ethyl acetate solution and were then correlated by the equation obtained by the slope of the calibration chart (**Figure 5**).

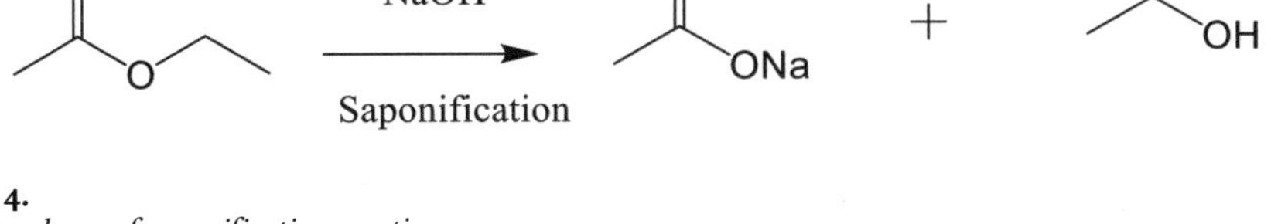

Figure 4.
Reaction scheme of saponification reaction.

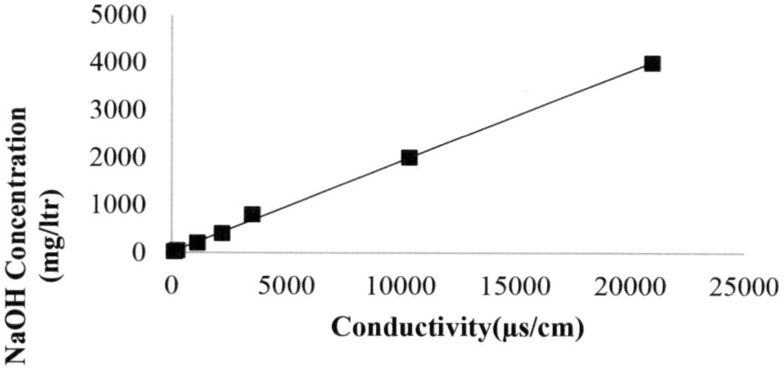

Figure 5.
Calibration chart of standard concentration of NaOH and its conductivity values.

4.1 RTD studies

The experiments were performed as per the procedure [22] for reactor type 1 and 2. A constant flow rate of 15.8 mL/min was set for all the tracer studies. The pulse input injection was done by the construction of a T-joint and an injection port, where a 10 mL tracer was used for every injection. For step input studies the input to the pump was transferred from inline water flow into the beaker containing the tracer solution. As we proceeded with the characterization of the flow reactors, the pulse method of injection was found to generate inaccurate and inconsistent data. So, final quantification experiments were conducted only with step inputs. Apart from sodium hydroxide, Rhodamine B was also used for the detection of flow patterns and presented data (**Table 1**).

The C, F, and E curves were drawn for each of the reactors set up. **Figures 6–8** represents curves for the tubular reactor of type 1. We observed an $F(t)$ of 3.5 minutes was 0.12, 5.5 minutes was 0.35, which means 12% and 35% of the molecules spent less the 3.5 and 5.5 minutes respectively in the reactor. Also, we could derive 80% of the molecules spend 13 minutes or less in the reactor and around 20% of the molecules spend longer than 13 min in the reactor. We find that around 54% of the material leaving the reactor spends between 3.5 and 5.5 minutes. The long-time portion in this is between 21 and 36 minutes, which accounts for 3% of the material being spent in the reactor.

Type of reactor	Mean residence time, t_m, s	Variance, σ^2, s^2	Skewness, S, s^3	Dispersion number, D/uL	Reynolds number, Re	Bodenstein number, B_0	Damköhler number, Da
Tubular reactor (type 1)	543.19	124,890	7.70E+07	0.00068	47.58	1453	0.163
Packed bed reactor (type 2)	676.8	244,385	1.37E+08	0.0122	0.215	81.35	1.57

Table 1.
Experimental residence time distribution parameters for reactor type 1 and 2.

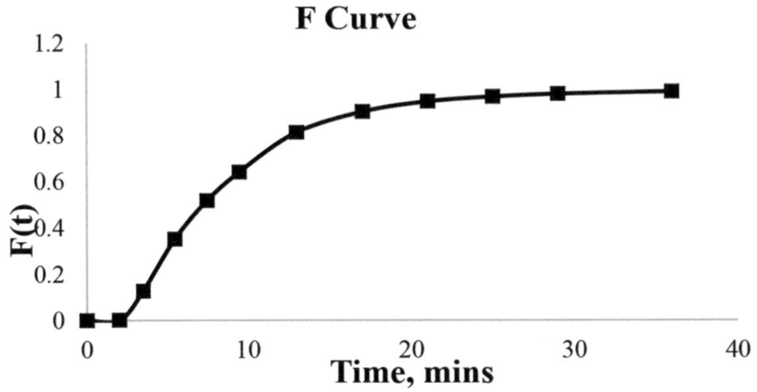

Figure 6.
F-curve, change in concentration (C/C_o) with respect to time.

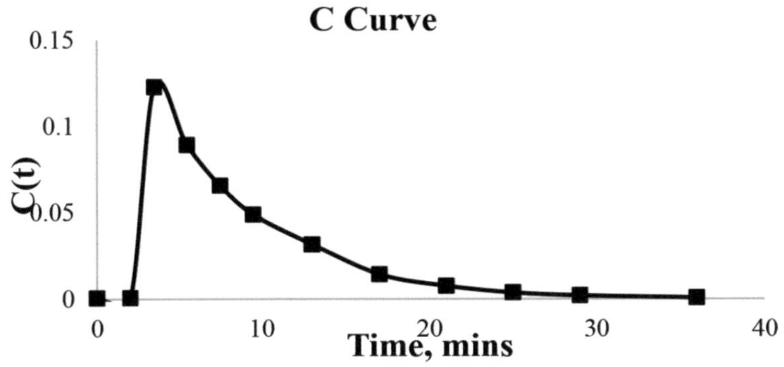

Figure 7.
C-curve, tracer concentration with respect to time.

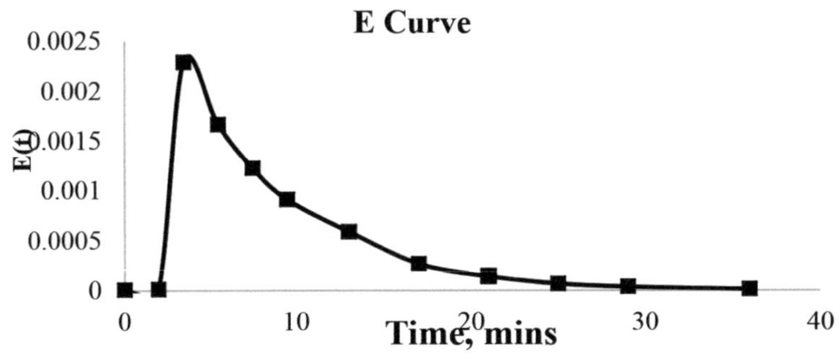

Figure 8.
E-curve, tracer concentration with respect to time.

Similarly, the C, F, and E curves were developed for the packed bed reactor (type 2). **Figures 9–11** represent curves for the type 2 reactor setup. Here we observed an $F(t)$ of 4 minutes was 0.21, 6 minutes was 0.35, which means 21% and 35% of the molecules spent less the 4 and 6 minutes respectively in the reactor. Also, we could derive 81% of the molecules spend 10 minutes or less in the reactor and around 20% of the molecules spend longer than 14 min in the reactor. We find that around 42% of the

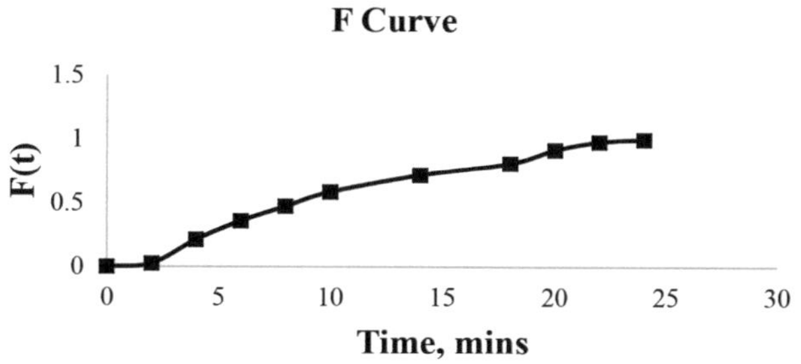

Figure 9.
F-curve, change in concentration (C/C_0) with respect to time.

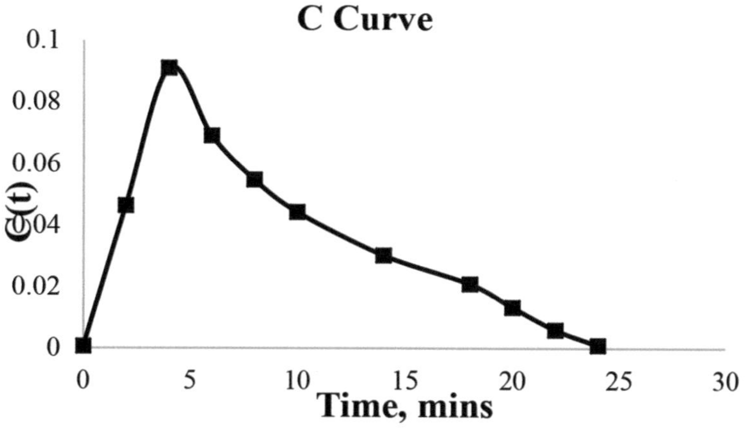

Figure 10.
C-curve, tracer concentration with respect to time.

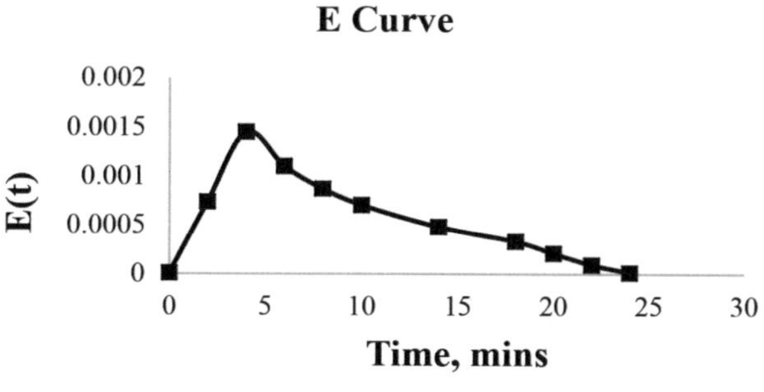

Figure 11.
E-curve, tracer concentration with respect to time.

material leaving the reactor spends between 4 to 6 minutes. The long-time portion in this is between 18 and 24 minutes, which accounts for 5.5% of the material being spent in the reactor. So, the type 1 reactor behavior is better in comparison with type 2 reactor concerning the performance and behavior close to the ideality. The performance curves for type 3 reactor are similar to the type 1 reactor, as both are tubular flow reactors. Additional experiments for type 2 reactor were performed under gravity and against gravity conditions to compare the behavior of the two methods. The trend clearly shows the deviation is quite large with gravity conditions due to by-passing, and channeling effects. To improve the performance, one could use running the packed columns against gravity with an add-on sintered plate at the top of the reactor which could change the dynamics and distributions of flow across the packed bed.

The mean residence time for type 1 and 2 reactors were found to be 9.05 and 11.28 minutes, where the average residence time was around 1.26 and 3.05 minutes respectively. This indicates there is a dispersion all along the fluid path and across boundaries.

The variance (σ^2) is an indication of the "spread" of the distribution; the greater the value, the higher the distribution across the path for the reactor. In our case, it was around 124,890 and 244,385 s^2, which clearly states that spread is almost twice in a packed bed reactor in comparison with the tubular reactor. The skewness factor measures the extent that a distribution is skewed in one direction or another about the mean. In general, all the three parameters mentioned above are essential to characterize the distribution and are enough to understand. A skewness factor(S) of around 7.70E+07 and 1.37E+08 sec^3 was found for type 1 and type 2 reactors respectively.

The dispersion numbers were also estimated a trial and error basis. First assume a small dispersion, say $\sigma^2/t_m{}^2$ is equal to 2D/uL and equate to get the appropriate dispersion. In our case, we found 0.00068 and 0.0122 for type 1 and type 2 reactors respectively, which clearly states a smaller dispersion in the case of type 1 and reasonable large dispersion in type 2 reactor from ideal plug flow behavior. Since the dispersion number varies along the length of the reactor and the mean residence time is higher than that of the theoretical residence time, the reactors are classified as closed systems. The changes in the packing materials used in the packed column had a significant effect on the flow pattern. To further simplify the plug flow behavior, the ratio of the length of the reactor and their effective diameters revealed to be higher than 50 for all the systems.

The Peclet number (Pe) was estimated and found to be 1470 and 81 for type 1 and type 2 reactors. The behavior of reactor type 1 is more or like the plug flow and reactor type 2 is behaving far from ideal conditions.

The Reynolds number (Re) is dimensionless describes the ratio of inertial to viscous forces. The regime for flow through a packed bed may be identified by the packed bed Re. The type 1 reactor falls under the transitional region and type 2 reactor falls under the laminar region.

Bodenstein number (B_0) was estimated and found to be 1452 and 81 for type 1 and type 2 reactor respectively. It could be concluded that both the reactors have varying degrees of back mixing, however, the variation in the flow velocity could be used to control or adjust B_0 for the desired condition.

Similarly, the Damköhler number (Da) was estimated to realize the mass transfer rates using the standard equation available in the literature [19]. A 0.163 and 1.57 were found for type 1 and 2 reactors, which signifies diffusion occurs much faster than the reaction, thus diffusion reaches equilibrium well before the reaction is at equilibrium for type 1 reactor and diffusion-limited system for type 2 reactor.

4.2 Estimation of kinetic parameters

Kinetic experiments were carried out both in batch and flow reactors type 1 and type 2 with defined procedures. The batch experiments were conducted first with concentrations of 0.02 N of ethyl acetate and sodium hydroxide, volumes of 100 mL each. Experiments were performed at three different temperatures such as 26.5, 33 and 44°C respectively. An estimated quantity of ethyl acetate was charged to the reactor and the desired temperature was set through the circulator. An agitation of around 300 RPM was set using the overhead motor connected to the reactor. Once the temperature is stable, the calculated volume of freshly prepared sodium hydroxide solution was dosed into the reactor at one shot, simultaneously the stopwatch was started. A change in the conductivity was noted over time. As the reaction progresses, the conductivity value will decrease, like sodium hydroxide being used in the reaction to form sodium acetate in the solution. The decrease in the concentration of sodium hydroxide was measured against conductance. The C_0, C_t, and C_∞ are the specific conductance of reaction mixtures at time zero, t, and infinity. Since the reaction follows second order kinetics, a plot of $(C_0 - C_t)/(C_t - C_\infty)$ versus time (**Figure 12**) was drawn to estimate rate constant (k) using the reliable method suggested in the literature [23]. Reaction conversion (X) was estimated using second-order kinetics, $X = (1 - C_A/C_{A0})$ at every time interval, and reported. The reaction conversion was of the order 68% (42 min), 67% (40 min) and 78% (203 min) for temperatures 26.5, 33 and 44°C. A plot of ln k versus T (**Figure 13**) and ln k/T versus 1/T was plotted (**Figure 14**) to estimate various thermodynamic parameters such as activation energy, activation enthalpy, activation entropy, and Gibbs free energy of activation. The k values (**Table 2**) for temperatures 26.5, 33 and 44°C were in the order of 0.14, 0.215, and 0.305 $Lmol^{-1}\ s^{-1}$ respectively.

A plot of ln k versus 1/T was plotted (**Figure 13** and **Table 3**) to estimate activation energy, the slope of the curve is −4118.5 and intercept of 11.837 to yield activation energy of 34.24 kJ/mol and Arrhenius constant of $8.05 \times 10^6\ min^{-1}$ using the formula

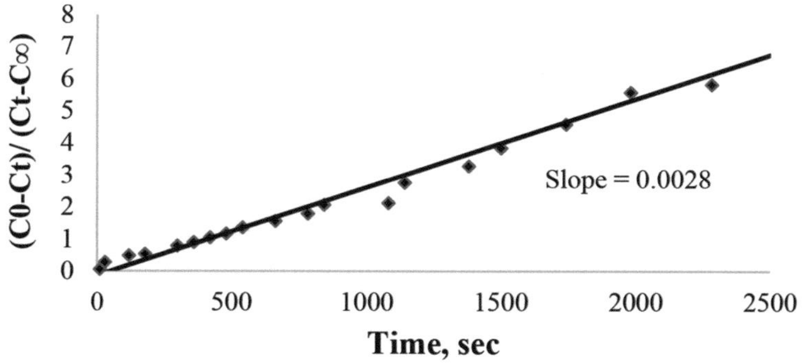

Figure 12.
Graphical determination of reaction rate constant.

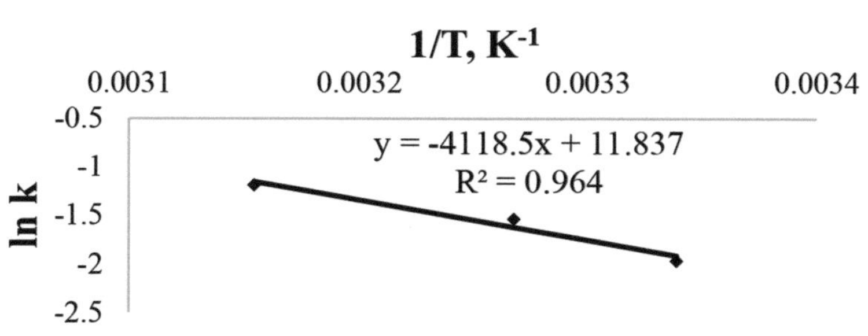

Figure 13.
Arrhenius plot of activation energy.

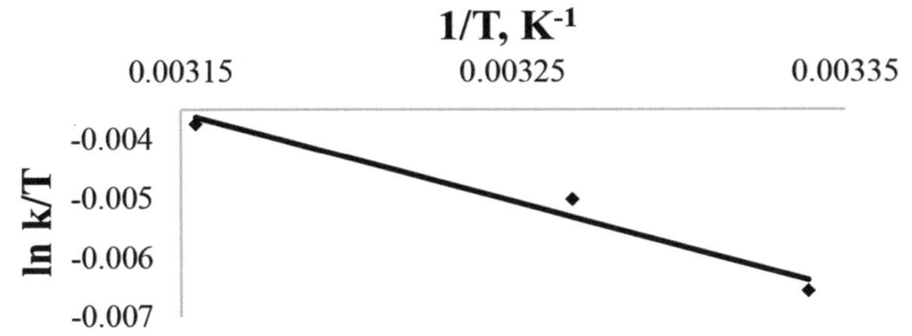

Figure 14.
Graphical evaluation of thermodynamic properties.

Temperature, K	1/T, K-1	Reaction rate constant, L/mol/s	ln k	ln k/T
299.5	0.003339	0.14	−1.96611	−0.00656
306	0.003268	0.215	−1.53712	−0.00502
317	0.003155	0.305	−1.18744	−0.00375
Slope of graph of ln k/T verses 1/T = −14.918				
Intercept of ln k/T verses 1/T = 0.0434				
Activation enthalpy: ΔH = −124.02 kJ/mol				
Activation entropy: ΔS = 197.18 JK^{-1} mol^{-1}				
Free energy of activation: ΔG = 65.24 kJ/mol				

Table 2.
Thermodynamic parameters of saponification reaction.

Temperature, °C	Temperature, K	1/T, K^{-1}	Reaction rate constant, L/mol/s	ln k
26.5	299.5	0.00334	0.14	−1.966
33	306	0.00327	0.215	−1.537
44	317	0.00315	0.305	−1.187
The slope of the graph of ln k versus 1/T = −4118.5 K				
Activation energy: E_a = 34.24 kJ/mol				

Table 3.
Effect of temperature on saponification reaction under batch conditions.

$k = Ae{-}Ea/RT$, where k = rate constant at temperature T, Ea = activation energy, R = universal gas constant and A = Arrhenius constant. The other thermodynamic parameters were estimated using the Eyring-Polanyi Equation [23], through plotting ln k/T versus 1/T (**Figure 14**). The slope of the curve was −14.918 and the intercept was 0.0434, from the slope the activation enthalpy was estimated as −124.02 kJ/mol, activation entropy of 197.18 JK^{-1} mol^{-1} and Gibbs free energy of activation was 65.24 kJ/mol (**Table 2**). The results of the rate constant and activation energy for saponification reaction are in good agreement and comparable with the literature [22, 24–27, 29]. There could be reasonable errors associated with the sensitivity of conductivity probe and methods of estimation reported by various authors reported in the literature [22].

Saponification experiments were conducted in the tubular reactor (type 1) of diameter 1.58 and 3.175 mm with a total reactor volume of 33.5 mL and varying flow rates from 6.7 and 13.4 mL/min respectively. Experiments were performed at two different temperatures such as 26.5 and 44°C. The concentrations of 0.02 N of ethyl acetate and sodium hydroxide solutions were used with varying residence time from 2.5, 5, and 10 min respectively. Once the temperature is stable in the thermostat, two pumps A & B were switched on to initiate the flow of sodium hydroxide solution and ethyl acetate solution into the tubular reactor, which was pre-calibrated for known residence time. Samples were collected regularly at the exit of the tube to measure the conductivity. Estimation of reaction conversion(X) was obtained using the standard irreversible bimolecular second-order equation, $X = (1 - (C_A/C_{A0}))$, and rate constants were estimated using equation $X_A/(1 - X_A) = k\,C_{A0}\,t$ individually and averaged across experiments. The conversions are in the order of 68–76.6% for experiments at 26.5 and 44°C respectively (**Figure 15**). The k values for temperatures 26.5 and 44°C were in the order of 0.347 and 0.419 $Lmol^{-1}\,s^{-1}$ respectively.

Further saponification experiments were conducted in a packed bed reactor (type 2) of diameter and length of the reactor as 24 and 240 mm respectively. The total volume of the reactor is around 105 mL with packing and 56 mL as an available volume for the reaction. The experiments were conducted under gravity and against gravity flow to check the performance of the reactor.

The flow rates for gravity flow experiments were in the range of 6.7, 7.8, and 8.4 mL/min respectively. Experiments were performed at two different temperatures such as 26.5 and 44°C. The concentrations of 0.02 N of ethyl acetate and sodium hydroxide solutions were used with varying residence time from 6.66, 7.17, and 8.35 min respectively. Once the temperature is stable in the thermostat, two pumps A & B were switched on to initiate the flow of sodium hydroxide solution and ethyl acetate solution into the reactor, which was pre-calibrated for known residence time. Samples were collected regularly at the exit of the tube to measure the conductivity. Estimation of reaction conversion (X) was obtained using a standard irreversible bimolecular second-order equation, $X = (1 - (C_A/C_{A0}))$, and rate constants were estimated using equation $X_A/(1 - X_A) = k\,C_{A0}\,t$ individually and averaged across experiments. The conversions are in the order of 56–63% for experiments at 26.5 and 44°C respectively (**Figure 16**). The k values for temperatures 26.5 and 44°C were in the order of 0.167 and 0.171 $Lmol^{-1}\,s^{-1}$ respectively.

Similar experiments were conducted in a packed bed reactor (type 2) against gravity flow through feeding the streams from the bottom of the reactor to minimize the channeling effect in the packed bed reactor.

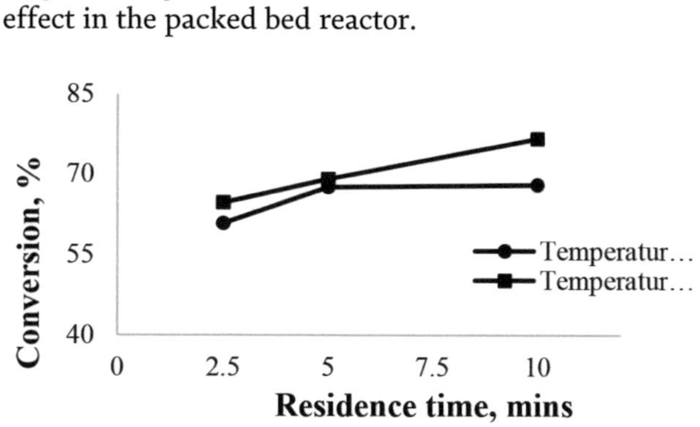

Figure 15.
A plot of reaction conversion verses residence time in tubular reactor (type-1).

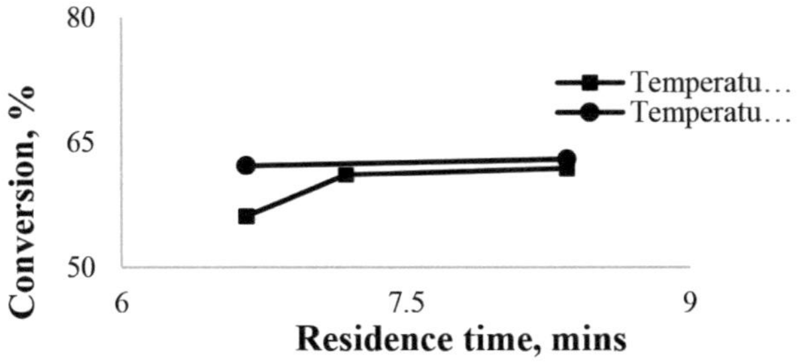

Figure 16.
A plot of reaction conversion verses residence time in packed bed reactor (type 2) under gravity flow.

The flow rates for gravity flow experiments were in the range of 6.8, 7.8, and 8.4 mL/min respectively. Experiments were performed at two different temperatures such as 26.5 and 44°C. The concentrations of 0.02 N of ethyl acetate and sodium hydroxide solutions were used with varying residence time from 6.66, 7.17, and 8.235 min respectively. Once the temperature is stable in the thermostat, two pumps A & B were switched on to initiate the flow of sodium hydroxide solution and ethyl acetate solution from the bottom of the reactor, which was pre-calibrated for known residence time. Samples were collected regularly at the exit of the tube to measure the conductivity. Estimation of reaction conversion (X) was obtained using a standard irreversible bimolecular second-order equation, $X = (1 - (C_A/C_{A0}))$, and rate constants were estimated using equation $X_A/(1 - X_A) = k\, C_{A0}\, t$ individually and averaged across experiments. The conversions are in the order of 63.7–73.3% for experiments at 26.5 and 44°C respectively (**Figure 17**). The k values for temperatures 26.5 and 44°C were in the order of 0.22 and 0.288 $Lmol^{-1}\ sec^{-1}$ respectively.

A new set of experiments was conducted in tubular bed reactor (type 1) submerged in Sonicator bath with sonication frequency (40 Hz) and performed experiments under the following conditions to check the effect of sonication on reaction kinetics under identical conditions.

The flow rates were in the range of 3.35, 6.7, and 13.4 mL/min respectively. Experiments were performed at two different temperatures such as 26.5 and 44°C. The concentrations of 0.02 N of ethyl acetate and sodium hydroxide solutions were used with varying residence time from 2.5, 5, and 10 min respectively. Once the temperature is stable, two pumps A & B were switched on to initiate the flow of sodium hydroxide solution and ethyl acetate solution to the reactor, which was pre-calibrated for known residence time. Samples were collected regularly at the exit of the tube to measure the conductivity. Estimation of reaction conversion(X) was obtained using a standard irreversible bimolecular second-order equation,

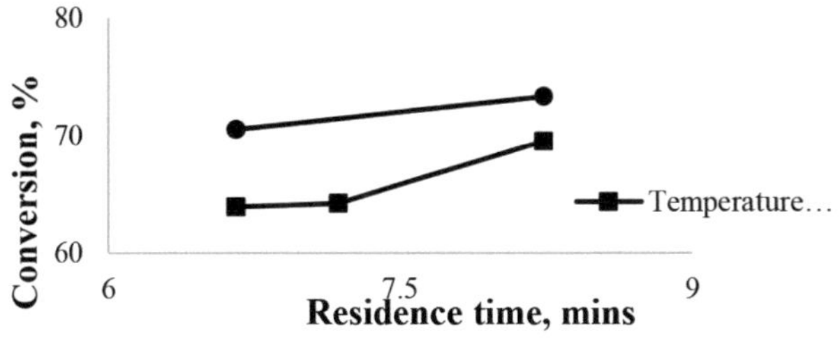

Figure 17.
A plot of reaction conversion verses residence time in packed bed reactor (type-2) under gravity flow.

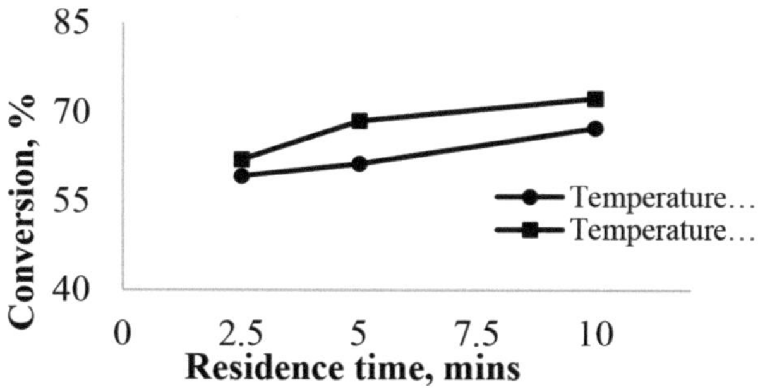

Figure 18.
A plot of reaction conversion verses residence time in tubular reactor (type-2) under sonication bath.

$X = (1 - (C_A/C_{A0}))$, and rate constants were estimated using equation $X_A/(1 - X_A) = k\ C_{A0}\ t$ individually and averaged across experiments. The conversions are in the order of 59–72.3% for experiments at 26.5 and 44°C respectively (**Figure 18**). The k values for temperatures 26.5 and 44°C were in the order of 0.30 and 0.374 $Lmol^{-1}\ s^{-1}$ respectively.

5. Conclusions

In the present work, customized flow reactors were designed, fabricated, and characterized through standard methodologies available in the literature. The characterization of the reactors was through RTD studies followed by estimation of various dimensionless parameters to understand the behavior such as E, F and C curves, mean residence time, variance, skewness factor, Reynolds number, Bodenstein number, Dispersion coefficients, and Damköhler number. The results have shown a noteworthy impact on these reactor designs, especially the tubular reactor and packed bed reactor under various operating regimes. The type 1 reactor falls under the transitional region and type 2 reactor falls under the laminar region. It could be concluded that both the reactors have varying degrees of back mixing observed across process conditions. In the case of dispersion, a smaller dispersion was found for type 1 and reasonably large dispersion in type 2 reactor, also the diffusion occurs much faster than the reaction, thus diffusion reaches equilibrium well before the reaction is at equilibrium for type 1 reactor and diffusion-limited system for type 2 reactor. Overall, the behavior of reactor type 1 is more or like the plug flow and reactor type 2 is behaving far from ideal conditions of the plug flow reactor. The spread of distribution was almost twice in a packed bed reactor in comparison with the tubular reactor.

Similarly, the reaction conversion across these reactors have shown significant variation across the reactor types under identical conditions. A significant reduction in residence time was observed in type 1 reactor versus type 2 and batch reactor to achieve similar or equivalent conversion. There is no much impact of ultrasonication on the reaction kinetics of type 1 reactor designs for the given reaction conditions. As a path forward, these reactors could be tested with other reactions or modify the design factors to understand the influence of design and operating conditions.

Author details

Girish Basavaraju[1,2*], Y. Ashwin[3], Prathyusha Naini[2] and Ravishankar Rajanna[1]

1 Department of Chemical Engineering, Dayananda Sagar College of Engineering, Bengaluru, India

2 Syngene International Ltd, Bengaluru, Karnataka, India

3 Manipal Institute of Technology, Manipal, Karnataka, India

*Address all correspondence to: girish.basavaraju@gmail.com

References

[1] Fell N, Ramshaw C. Innovation offers a new spin on drug production. Chemical Engineer (London). 1998;**656**: 23-25

[2] Gibbard I. Spinning disk reactors. New opportunities for the chemical industry. In: Proc. Intensif.: Profits for the Chem. Ind., Rotterdam. Sittard, The Netherlands: Netherlands Agency for Energy and the Env.; 1998

[3] Oxley P, Brechtelsbauer C, Ricard F, Lewis N, Ramshaw C. Evaluation of spinning disk reactor technology for the manufacture of pharmaceuticals. Industrial & Engineering Chemistry Research. 2000;**39**(7):2175-2182

[4] Jensen KF. Microreaction engineering—is small better? Chemical Engineering Science. 2001;**56**(2): 293-303

[5] Anderson NG. Practical use of continuous processing in developing and scaling up laboratory processes. Organic Process Research & Development. 2001; **5**(6):613-621

[6] Li CJ, Trost BM. Green chemistry for chemical synthesis. Proceedings of the National Academy of Sciences. 2008; **105**(36):13197-13202

[7] Souza JM, Galaverna R, Souza AAD, Brocksom TJ, Pastre JC, Souza ROD, et al. Impact of continuous flow chemistry in the synthesis of natural products and active pharmaceutical ingredients. Anais da Academia Brasileira de Ciências. 2018; **90**(1):1131-1174

[8] Klose F, Wolff T, Thomas S, Seidel-Morgenstern A. Concentration and residence time effects in packed bed membrane reactors. Catalysis Today. 2003;**82**(1-4):25-40

[9] Plugatyr A, Svishchev IM. Residence time distribution measurements and flow modeling in a supercritical water oxidation reactor: Application of transfer function concept. The Journal of Supercritical Fluids. 2008;**44**(1):31-39

[10] Carpenter NG, Roberts EPL. Mass transport and residence time characteristics of an oscillatory flow electrochemical reactor. Chemical Engineering Research and Design. 1999; **77**(3):212-217

[11] Bolin B, Rodhe H. A note on the concepts of age distribution and transit time in natural reservoirs. Tellus. 1973; **25**(1):58-62

[12] Garcia-Serna J, García-Verdugo E, Hyde JR, Fraga-Dubreuil J, Yan C, Poliakoff M, et al. Modelling residence time distribution in chemical reactors: A novel generalised n-laminar model: Application to supercritical CO_2 and subcritical water tubular reactors. The Journal of Supercritical Fluids. 2007; **41**(1):82-91

[13] Iliuta I, Thyrion FC, Muntean O, Giot M. Residence time distribution of the liquid in gas-liquid cocurrent upflow fixed-bed reactors. Chemical Engineering Science. 1996;**51**(20): 4579-4593

[14] Schwartz SE. Residence times in reservoirs under non-steady-state conditions: Application to atmospheric SO_2 and aerosol sulfate 1. Tellus. 1979;**31** (6):530-547

[15] Xi Y, Chen Q, You C. Flow characteristics of biomass particles in a horizontal stirred bed reactor: Part I. Experimental measurements of residence time distribution. Powder Technology. 2015;**269**:577-584

[16] Mata-Segreda JF. Hydroxide as general base in the saponification of ethyl acetate. Journal of the American Chemical Society. 2002;**124**(10): 2259-2262

[17] Levenspiel O. Chemical reaction engineering. Industrial & Engineering Chemistry Research. 1999;**38**(11): 4140-4143

[18] Pinder KL. Fogler, HS "elements of chemical reaction engineering", prentice-hall, englewood cliffs, New Jersey 07632, 1986, 769 pages. Price $74.15 Canadian. The Canadian Journal of Chemical Engineering. 1987;**65**(3): 526-527

[19] Wen CY, Fan LT. Models for Flow Systems and Chemical Reactors. M. Dekker; 1975. ISBN: 0824763467. Available from: https://agris.fao.org/agris-search/search.do?recordID=US201300516754

[20] Ravi R. Flow characteristics of reactors—Flow modeling. In: Coulson and Richardson's Chemical Engineering. Butterworth-Heinemann; 2017. pp. 103-160. ISBN: 978-0-08-101096-9. Available from: https://www.sciencedirect.com/book/9780081010969/coulson-and-richardsons-chemical-engineering#book-info

[21] Coulson JM, Backhurst JR, Richardson JF, Harker JH. Chemical Engineering: Solutions to the Problems in Chemical Engineering. Vol. 5. Pergamon; 1979. ISBN 13: 9780080229522

[22] Das K, Sahoo P, Sai Baba M, Murali N, Swaminathan P. Kinetic studies on saponification of ethyl acetate using an innovative conductivity-monitoring instrument with a pulsating sensor. International Journal of Chemical Kinetics. 2011;**43**(11):648-656

[23] Sharma LR, Pathania MS, Puri BR. Principles of Physical Chemistry. Vishal publishing; 2010

[24] Danish M, Al Mesfer MK. A comparative study of saponification reaction in a PFR and CSTR. Research Journal of Chemical Sciences. 2015;**5**(11): 13-17

[25] Tsujikawa H, Inoue H. The reaction rate of the alkaline hydrolysis of ethyl acetate. Bulletin of the Chemical Society of Japan. 1966;**39**(9):1837-1842

[26] Patil DB, Batra V, Kapoor SB. Kinetic studies on saponification of poly (ethylene terephthalate) waste powder using conductivity measurements. Journal of Polymers. 2014;**2014**:1-7. DOI: 10.1155/2014/321560

[27] Jensen FW, Watson GM, Beckham JB. Study of saponification reaction rate of ethyl acetate by high frequency titrimetry. Analytical Chemistry. 1951;**23**(12):1770-1773

[28] Bursali N, Ertunc S, Akay B. Process improvement approach to the saponification reaction by using statistical experimental design. Chemical Engineering and Processing: Process Intensification. 2006;**45**(11):980-989

[29] Warder RB. Relation between temperature and the rate of chemical reactions. American Chemical Journal. 1881;**3**:203-204

Chapter 3

Catalytic Behavior of Extended π-Conjugation in the Kinetics of Sensitizer-Mediator Interaction

Rozina Khattak

Abstract

This chapter discusses the catalytic effect of extended π-conjugation on the electron transfer process between ferricyphen-ferrocyanide and ferricypyr-ferrocyanide in an aqueous medium. Ferricyphen and ferricypyr may be feasible options for the sensitizer in dye-sensitized solar cells due to their high reduction potential, stability, capability as an outer-sphere oxidant, and photosensitivity. Meanwhile, ferrocyanide could be used as a mediator in DSSCs instead of iodide to avoid iodate production and achieve a similar reduction potential and stability. This chapter compared the ability of competent putative sensitizers to oxidize the likely mediator in water. In contrast to the 2,2′-dipyridyl chelate, the extended π-conjugation in 1,10-phenanthroline accelerated the redox process by increasing the electron affinity of ferricyphen as compared to ferricypyr. The reactions had the same kinetics but different rate constants, indicating that the ferricyphen-ferrocyanide reaction was several times faster than the ferricypyr-ferrocyanide reaction, revealing and confirming the catalytic influence of extended π-conjugation on the redox process.

Keywords: pi-conjugation, ferricyphen, ferricypyr, ferrocyanide, catalysis, kinetics, aqueous medium

1. Introduction

A dye-sensitized solar cell (DSSC) relies on the interaction of the sensitizer and the flow of electrons from the mediator to the sensitizer to complete the circuit and convert photo energy to electrical energy. As a result, the sensitizer-mediator interaction in every DSSC is critical to its efficiency and stability. The reduced and oxidized forms of the sensitizer, as well as the reduced and oxidized forms of the mediator, must have adequate stability in the reaction medium to ensure a successful electron-transfer process. Meanwhile, the solubility of the oxidized and reduced sensitizers and mediators is an important metric to consider while building a DSSC. A photoanode, sensitizer, mediator, solvent, and the counter electrode are all common components of a DSSC, and the electrolyte is sometimes employed as well.

Scientists and engineers prefer DSSCs over other first and second-generation solar cells, such as thin-film and silicon-based solar cells, since they are less expensive, more stable, and environmentally friendly. Because DSSCs' maximum efficiency is lower than that of first and second-generation solar cells, which is 12–14% under ideal conditions of materials and structure using Ru(II) dyes; ruthenium in the 2+ oxidation state, compared to 20–30% for the latter two types of cells, increasing their efficiency in a cost-effective and environmentally benign manner is still a hot topic. As a result, there is still a lot to learn about DSSCs and how to increase their efficiency, stability, and longevity by employing better conditions, materials, and structures.

Researchers have examined a number of materials for the photoanode and counter electrode to improve the efficacy of a DSSC, including nanomaterials, their composites, and nanocomposites of the same materials with varied morphologies [1–5]. In this investigation, heteroatom-doped graphene catalysts were used [6]. Heterogeneous $FeNi_3/NiFe_2O_4$ nanoparticles with modified graphene were also explored as electrocatalysts for dye-sensitized solar cells [7]. Studies on making ecologically acceptable and stable DSSCs using natural dyes and NHC-based iron as sensitizers [8–10], copper and other transition metal-based mediators [11, 12], and electrolytes based on gels and polymers [13–17] have been published. The steric, structural, and compositional effects of solvent on the efficiency of DSSC have also been investigated electrochemically [15, 18].

The use of potential Fe-based sensitizers and mediators has been revealed in this chapter based on their comparative kinetic study. The rate of the electron-transfer reactions is influenced by the structure of the sensitizer that has been discussed in this chapter. Iron-based sensitizers are less expensive and environmentally benign than ruthenium-based sensitizers, making them attractive to the socio-economic impact. The solubility of iron-based sensitizers in an aqueous medium, as well as the use of an aqueous medium rather than inflammable, volatile, poisonous, and expensive chemical solvents, are two further advantages. Iodate is produced by the electrolytic solution of iodide/triiodide, which is corrosive to stainless steel and a source of DSSC instability. Another approach to a stable DSSC is the use of an iron-based coordination complex such as ferrocyanide as a mediator that has one electron transfer chemistry and comparable redox potential to iodide/triiodide electrolyte. The one-electron transfer chemistry helps to reduce the recombination losses.

The effect of structure, such as extended π-conjugation, on the rate of the electrontransfer between the sensitizer-mediator in an aqueous medium, is the discussion in this chapter. Ferricyphen and ferricypyr are the names for dicyanobis(1,10-phenanthroline)iron(III) and dicyanobis(2,2′-dipyridyl)iron(III), respectively, where "ferri" stands for Fe(III) oxidation state, "cy" for cyanide, and "phen" or "pyr" for the chelate. Reduced variants are known as ferrocyphen and ferrocypyr. Both ferricyphen and ferricypyr are substitution inert outer sphere oxidants with octahedral geometry and similar Fe(III) transition metal coordination sites (**Figure 1**). Their reduction potentials are 0.80 V and 0.76 V, respectively, however, they were initially synthesized in the 1960s [19–22]. They are potential sensitizers because of their photosensitive nature and their solubility in an aqueous medium in the oxidized form and comparatively low solubility in the reduced form which may be helpful for their adsorption on the photoanode. Each of the potential sensitizers easily oxidizes the selected potential mediator such as ferrocyanide in an aqueous medium without the need for any external triggering to initiate the reaction. Each of the redox reactions starts spontaneously after mixing the aqueous solutions of both reactants such as ferricyphen-ferrocyanide or ferricypyr-ferrocyanide. The reduction potential of ferrocyanide is comparable to the iodide electrolyte, hence displays its replacement over iodide. The oxidation of iodide by ferricyphen and ferricypyr has been studied in

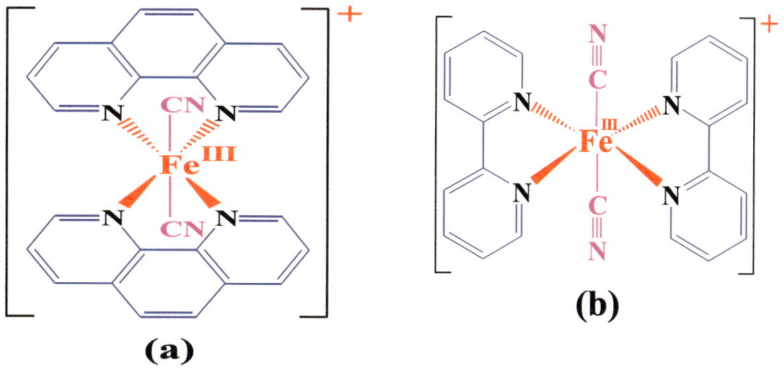

Figure 1.
(a) Dicyanobis(1,10-phenanthroline)iron(III). (b) Dicyanobis(2,2′-dipyridyl)iron(III).

acetonitrile, aqueous tertiary butyl alcohol, and aqueous 1,4-dioxane [23–26]. The comparative kinetic analysis shows the rapid kinetics of ferricyphen over ferricypyr in binary solvent media under optimized experimental conditions. The following section of the chapter will help to identify the role of pi-conjugation in reaction kinetics.

2. Methodology and kinetics studies

Kinetics of the reduction of ferricyphen, and or, ferricypyr (oxidizing agent) was studied in an aqueous medium under the pseudo-first-order condition. The concentration of ferrocyanide (reducing agent) was always in excess of the oxidizing agents. The reactions were probed at room temperature i.e., 25 °C, and at constant ionic strength i.e., 0.06 M. The concentration ratio between the oxidizing agents and the reducing agent was always maintained at 1:2.5, 1:5, 1:7.5, and 1:10, respectively. The reactions were probed spectrophotometrically under ordinary experimental conditions. No specific or extraordinary experimental setup was required such as an inert atmosphere, dark room, and/or a catalyst. However, the fresh solutions of the reactants were prepared and wrapped in aluminum foil soon after preparation because ferricyphen and or ferricypyr get reduced when their aqueous solutions are exposed to light. The reactions were started upon mixing the reactants and a rise in the absorbance was monitored as a function of time (**Figure 2A**). The instrumental setup consisted of a home-built assembly as mentioned earlier [27]. The molar absorptivity of the reduced ferricyphen i.e., ferrocyphen, and reduced ferricypyr i.e., ferrocypyr are several folds higher than the oxidized ferrocyphen (ferricyphen) and ferrocypyr (ferricypyr). The spectra of the reactants and products are shown in **Figure 2B** and compared to the literature [21–29] that supports the electron transfer mechanism. The integration method was implemented to figure out the rate constant. Each experiment was repeated three to six times for accuracy and the rate constant is an average value. The reactions were kinetically examined, and it was determined that each one among them was completed in two phases.

The rate of both reactions was found independent of the concentration of the oxidizing agents i.e., sensitizers, and the reducing agent i.e., mediator during the first phase of the reaction. The rate of the second phase of both reactions, on the other hand, was shown to be first order and dependent on the concentrations of the sensitizers and mediator. The linear rate equations (integration method) of zero and first

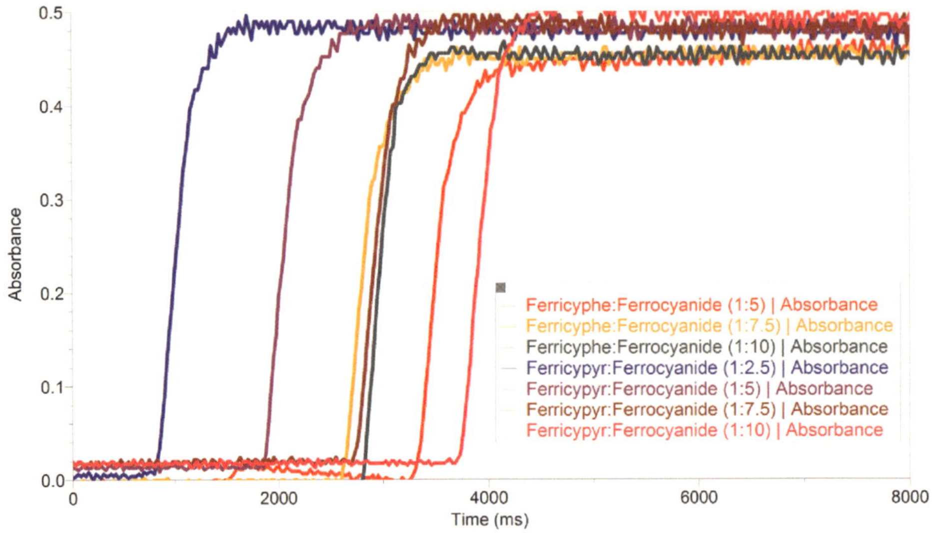

(A)

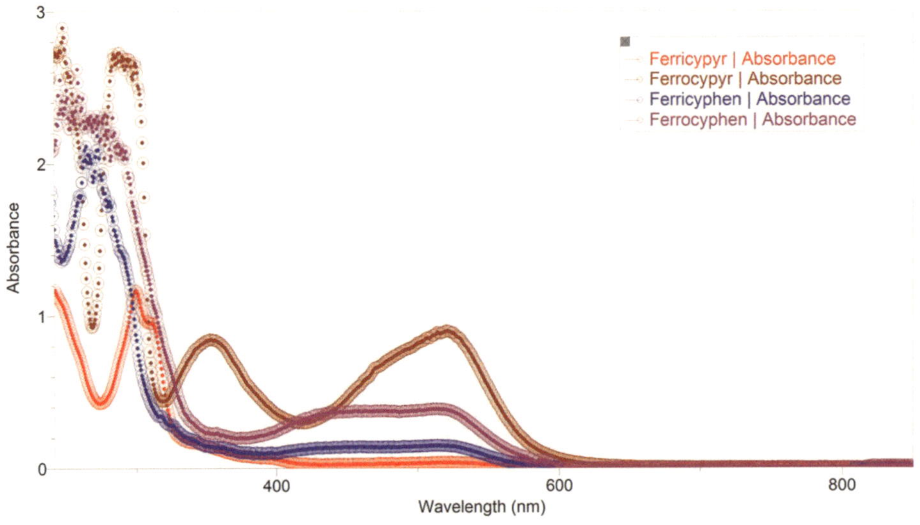

(B)

Figure 2.
(A) Time course graphs at a varying concentration of ferrocyanide and fixed ferricyphen/ferricypyr.
(B) UV-visible absorption spectral analysis of the sensitizer-mediator redox reactions.

order were best fitted with the highest linear fit R^2 value on the time course data in the first and the second phases of the reactions. The slope of each plot yielded the observed zero order rate constant (k_{obs}) and the observed pseudo-first-order rate constant (k'_{obs}), respectively. The effect of variation in the concentration of each of the sensitizers and the mediator on each of the rate constant was studied. It is assumed that if the concentration of the reactant is low and is varied, the rate constant should not be varied rather the rate of the reaction is varied according to the Eqs. (1)–(5) considering the pseudo-first-order condition. However, the rate constant is varied when the concentration of the reactant that has been taken in excess is varied rather than the rate of the reaction in case of the first order dependence on the mediator. Similarly, if it is zero order, the rate constant will have no effect by variation in the concentration of the mediator. **Figure 3** depicts the outcomes.

$$Rate = k_{obs}[Sensitizer]^{0}\ first\ phase\ of\ the\ reaction \tag{1}$$

$$Rate = k_{obs} \tag{2}$$

Since sensitizer << mediator

$$\therefore k_{obs} = k_{1}[Mediator]^{0}\ first\ phase\ of\ the\ reaction \tag{3}$$

$$k_{obs} = k_{1} \tag{4}$$

$$Rate = k'_{obs}[Sensitizer]\ second\ phase\ of\ the\ reaction \tag{5}$$

Since sensitizer << mediator

$$\therefore k'_{obs} = k_{2}[Mediator]\ second\ phase\ ofthe\ reaction \tag{6}$$

Eqs. (4) and (6) reveal k_1 as the overall zero order rate constant of the first phase of the reactions and k_2 as the overall second order rate constant of the second phase of the reactions. A first order is observed corresponding to the concentration of the mediator in each case i.e., ferricyphen/ferrocyanide phase-II and ferricypyr/ferrocyanide phase-II (**Figure 3**). The plots have intercepts that interpret the initial zero order reaction phase corresponding to the concentration of the mediator. However, the rest of the plots (**Figure 3**) reveal the results according to the Eqs. (1)–(5). The rate constants either observed zero order rate constant (k_{obs}) and the observed pseudo-first order rate constant (k'_{obs}) were independent of the concentration of the sensitizers in both phases and the mediator in the first phase of the reaction. Because of the low concentration that was maintained to follow the pseudo-first order condition, the rate constants corresponding to the sensitizers were independent of the concentration terms of the sensitizers in both phases. As a result, the findings show that the

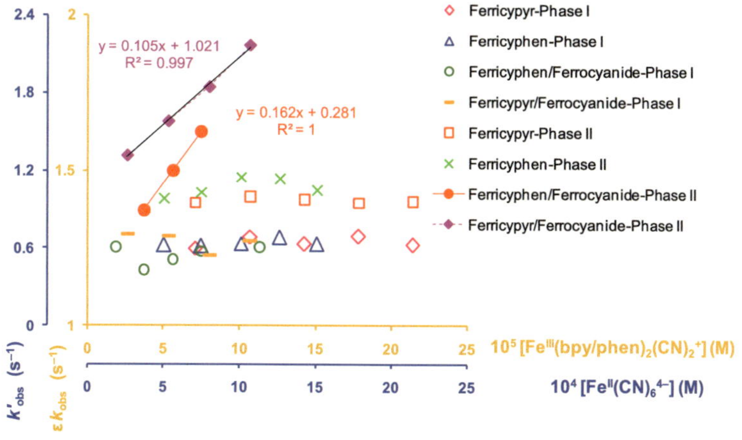

Figure 3.
Kinetic analysis of the sensitizer-mediator reactions.

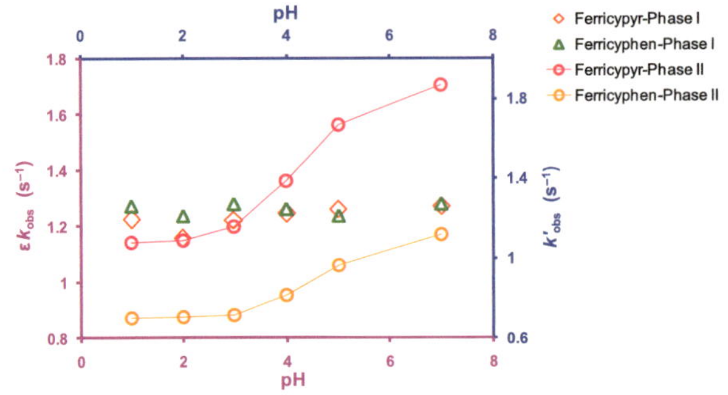

Figure 4.
Effect of pH on the rate constants of the first and second phases of each sensitizer-mediator reaction.

pseudo-first order criterion was successfully implemented. It is worth noting that the first phase of both reactions was long enough to get the reactions to about 70% completion. In the first phase of both of the reactions, the zero order rate constant (k_{obs}) is the multiplication product with the molar absorptivity (ε) of either of ferrocyphen or ferrocypyr, respectively. The zero order integrated rate equation (linear-fit) was implemented on the absorbance data rather concentration data of the time course graphs. Therefore, the slope of the plot was the multiplication product of $\varepsilon \bullet k_{obs}$ at the pathlength of the quartz cuvette equal to 1 cm. This multiplication of the constant value to the rate constant just adds a constant mathematical figure to the rate constant and does not affect the rate constant and overall findings of the data and the results.

For further rectification of the results, the effect of pH was monitored on the rate constants in each phase of the reactions under the pseudo-first order condition. The concentration of the mediator was always in excess over the sensitizers and the concentration of the nitric acid was always in excess over the mediator at room temperature and constant ionic strength 0.12 M. The results are revealed in **Figure 4** by plotting the graphs between pH and the rate constants on x-y coordinates respectively for each of the sensitizer-mediator interaction. The first phase of each reaction was observed unaffected of pH that declares and confirms the zero order reaction in this phase of each reaction. However, the second phase of the reaction shows curvatures (**Figure 4**). The value of the pseudo-first order rate constant (k'_{obs}) decreased with decreasing pH and became constant at the low values of the pH as has been shown in the **Figure 4**. These results indicate the formation of the monoprotonated ferrocyanide upon increasing the acidity of the reaction medium via conversion of free ferrocyanide species to the rate-inhibiting monoprotonated ferrocyanide and though the value of the rate constant decreased in each case. The protonation of the sensitizers under the pH employed have not been mentioned in the literature [22]. However, ferrocyanide may form mono- to tetra-protonated species considering the charge on the free ferrocyanide and depending on the acidity of the reaction medium [30].

3. Catalytic effect of extended pi-conjugation

The comparative analysis of both of the sensitizer-mediator redox reactions revealed the catalytic effect of pi-conjugation on the rate of reaction. The second order rate constant (k_2) in case of ferricyphen-ferrocyanide has a greater value i.e.,

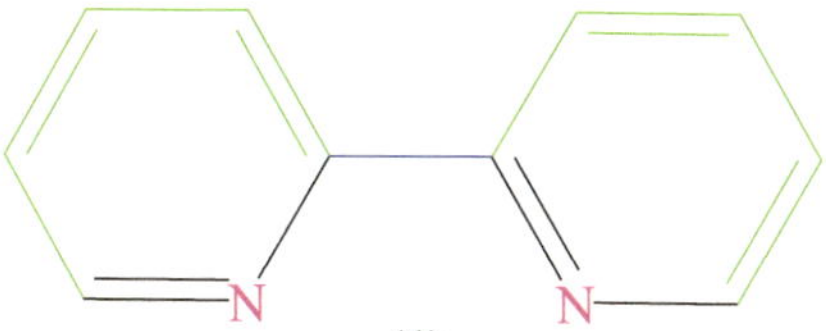

(1)

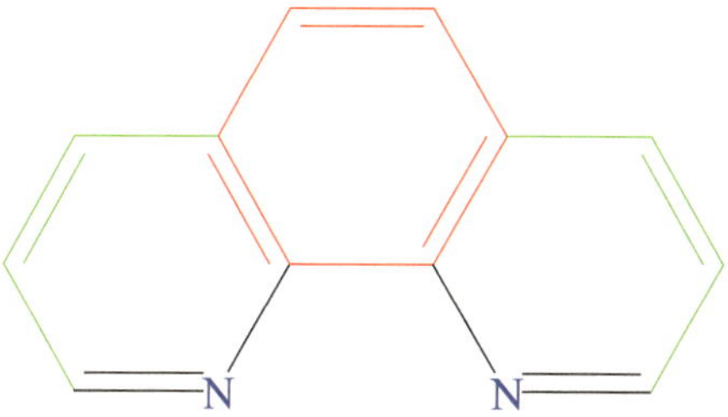

(2)

Figure 5.
(1) 2,2'-Dipyridyl. (2) 1,10-Phenanthroline.

1620 M^{-1} s^{-1} as compared to the ferricypyr-ferrocyanide that is 1050 M^{-1} s^{-1}. The kinetic analysis of both of the reactions helped to identify the similar reaction mechanism of electron transfer between each sensitizer-mediator pair with a high rate in case of ferricyphen-ferrocyanide as compared to ferricypyr-ferrocyanide under similar experimental conditions. The sensitizer such as ferricyphen has extended π-conjugation in its structure that distributes the electron density over the entire chelate molecule while stabilizing it. As a result, in comparison to the 2,2′-dipyridyl chelate, the availability of the lone pair of electrons on the nitrogen atoms in the 1,10-phenanthroline chelate is reduced (**Figure 5**). The chelate such as 2,2′-dipyridyl does not carry an extended π-conjugation in its structure though its nitrogen atoms have high density of the lone pair of electrons that can be coordinated by the Fe(III) atom of ferricypyr relatively easily. However, in instance of 1,10-phennathroline the extended π-conjugation of the structure reduces the density of the lone electrons on the nitrogen atoms and therefore availability to Fe(III) in ferricyphen. As a result, in order to circumvent this electron deficiency, the electron affinity of Fe(III) in ferricyphen increases as compared to ferricypyr, which helps to enhance its electron accepting potential as compared to ferricypyr. Consequently, a catalytic effect of extended π-conjugation is observed in the reduction kinetics of the ferricyphen as compared to ferricypyr by ferrocyanide in aqueous medium.

4. Conclusion

The kinetic study of the electron-transfer reaction between ferricyphen-ferrocyanide and ferricypyr-ferrocyanide revealed a complex mechanism. The reactions were completed into two phases. The initial phase of the reactions lasted long enough to complete the reaction at 70% efficiency and was independent of the sensitizer-mediator concentrations. This phase kinetics has been the most straightforward, with zero order corresponding to the sensitizer and mediator in the aqueous medium. The second phase of the reactions, on the other hand, was long enough to account for up to 30% of the total reaction time, with a first-order relationship of the rate of redox reactions on the concentration of sensitizers and mediator. In the second phase, the reactions followed an overall second order. The rate, that is second order rate constant, of the reaction was shown to be dependent on pH and the concentration terms involved, indicating that it was a rate-determining step involving interaction between ferricyphen or ferricypyr and ferrocyanide. However, as free ferrocyanide is consumed to generate monoprotonated ferrocyanide with increasing acidity, the rate of reaction in this phase slows down, indicating that this is the slow stage of the process. As a result, it has been discovered that monoprotonated ferrocyanide reduces sensitizer in the first phase of the reaction, which is a fast kinetic step in the electron transfer process. Meanwhile, the findings demonstrated that pi-conjugation in the sensitizer has a catalytic influence on the redox kinetics of the sensitizer-mediator interaction. When compared to less conjugated sensitizer that is ferricypyr, pi-conjugation increases the coordination compound's or sensitizer's (ferricyphen's) electron affinity, allowing it to receive electrons and oxidize the mediator more quickly.

Author details

Rozina Khattak
Department of Chemistry, Shaheed Benazir Bhutto Women University, Peshawar, Pakistan

*Address all correspondence to: rznkhattak@yahoo.com; rznkhattak@sbbwu.edu.pk

References

[1] Oh WC, Cho KY, Jung CH, Areerob Y. Hybrid of graphene based on quaternary $Cu_2ZnNiSe_4$-WO_3 nanorods for counter electrode in dye-sensitized solar cell application. Scientific Reports. 2020;**10**(1):4738

[2] Cavallo C, Di Pascasio F, Latini A, Bonomo M, Dini D. Nanostructured semiconductor materials for dye-sensitized solar cells. Journal of Nanomaterials. 2017;**2017**:5323164

[3] Raj Kumar T, Shaheer Akhtar M, Gnana Kumar G. Ni–Co bimetallic nanoparticles anchored reduced graphene oxide as an efficient counter electrode for the application of dye sensitized solar cells. Journal of Materials Science: Materials in Electronics. 2017; **28**(1):823-831

[4] Chen Y, Zhang H, Chen Y, Lin J. Study on carbon nanocomposite counterelectrode for dye-sensitized solar cells. Journal of Nanomaterials. 2012; **2012**:601736

[5] Samantaray MR, Mondal AK, Murugadoss G, Pitchaimuthu S, Das S, Bahru R, et al. Synergetic effects of hybrid carbon nanostructured counter electrodes for dye-sensitized solar cells: A review. Materials (Basel). 2020;**13**(12): 2779

[6] Zhao Z, Lin C-Y, Tang J, Xia Z. Catalytic mechanism and design principles for heteroatom-doped graphene catalysts in dye-sensitized solar cells. Nano Energy. 2018;**49**: 193-199

[7] Pang B, Zhang M, Zhou C, Dong H, Ma S, Feng J, et al. Heterogeneous $FeNi_3$/ $NiFe_2O_4$ nanoparticles with modified graphene as electrocatalysts for high performance dye-sensitized solar cells. Chemical Engineering Journal. 2021;**405**: 126944

[8] Ung MC, Sipaut CS, Dayou J, Liow KS, Kulip J, Mansa RF. Fruit based dye sensitized solar cells. IOP Conference Series: Materials Science and Engineering. 2017;**217**:012003

[9] Duchanois T, Liu L, Pastore M, Monari A, Cebrián C, Trolez Y, et al. NHC-based iron sensitizers for DSSCs. Inorganics. 2018;**6**(2):63, 10.3390/ inorganics6020063

[10] Li C-T, Lin RY-Y, Lin JT. Sensitizers for aqueous-based solar cells. Chemistry—An Asian Journal. 2017;**12**(5):486-496

[11] Bignozzi CA, Argazzi R, Boaretto R, Busatto E, Carli S, Ronconi F, et al. The role of transition metal complexes in dye sensitized solar devices. Coordination Chemistry Reviews. 2013;**257**(9): 1472-1492

[12] Yang K, Yang X, Zhang L, An J, Wang H, Deng Z. Copper redox mediators with alkoxy groups suppressing recombination for dye-sensitized solar cells. Electrochimica Acta. 2021;**368**:137564

[13] Dinari M, Momeni MM, Goudarzirad M. Dye-sensitized solar cells based on nanocomposite of polyaniline/ graphene quantum dots. Journal of Materials Science. 2016;**51**(6):2964-2971

[14] Dissanayake MAKL, Umair K, Senadeera GKR, Kumari JMKW. Effect of electrolyte conductivity, co-additives and mixed cation iodide salts on efficiency enhancement in dye sensitized solar cells with acetonitrile-free electrolyte. Journal of Photochemistry and Photobiology A: Chemistry. 2021; **415**:113308

[15] Kalaignan GP, Kang M-S, kang YS. Effects of compositions on properties of PEO–KI–I2 salts polymer electrolytes for DSSC. Solid State Ionics. 2006;**177**(11): 1091-1097

[16] Yu W-C, Lin L-Y, Chang W-C, Zhong S-H, Su C-C. Iodine-free nanocomposite gel electrolytes for quasi-solid-state dye-sensitized solar cells. Journal of Power Sources. 2018;**403**: 157-166

[17] Iftikhar H, Sonai GG, Hashmi SG, Nogueira AF, Lund PD. Progress on electrolytes development in dye-sensitized solar cells. Materials (Basel). 2019;**12**(12):1998

[18] Ramirez-Perez J, Maria C, Santacruz CP. Impact of solvents on the extraction and purification of vegetable dyes onto the efficiency for dye-sensitized solar cells. Renewables: Wind, Water, and Solar. 2019;(1):6, 1

[19] Wang X, Stanbury DM. Direct oxidation of l-cysteine by $[Fe^{III}(bpy)_2(CN)_2]^+$ and $[Fe^{III}(bpy)(CN)_4]^-$. Inorganic Chemistry. 2008; **47**(3):1224-1236

[20] Takagi HD, Kagayama N, Matsumoto M, Tarumi T, Funahashi S. Mechanistic study of oxidation reactions of hydroquinone, catechol and L-ascorbic acid by dicyanobis(1,10-phenanthroline)iron(III) in dimethyl sulfoxide. Journal of Molecular Liquids. 1995;**65-66**:277-280

[21] Khattak R. Comparative Kinetic Study for the Electron Transfer Reactions of Some Iron Complexes [PhD]. Karachi: University of Karachi; 2011

[22] Schilt AA. Mixed ligand complexes of iron(II) and (III) with cyanide and aromatic di-imines. Journal of the American Chemical Society. 1960; **82**(12):3000-3005

[23] Khattak R, Khan MS, Iqbal Z, Ullah R, Khan A, Summer S, et al. Catalytic effect of 1,4-dioxane on the kinetics of the oxidation of iodide by dicyanobis(bipyridine)iron(III) in water. Catalysts. 2021;**11**(7):840. DOI: 10.3390/catal11070840

[24] Wang X, Stanbury DM. Copper catalysis of the oxidation of iodide by $[Fe^{III}(bpy)_2(CN)_2]^+$ in acetonitrile. The Journal of Physical Chemistry. A. 2004; **108**(38):7637-7638

[25] Khattak R, Khan MS, Ullah R, Zainab S, Ali M, Rahman W, et al. Effect of the ionic strength on the redox reaction of dicyanobis(bipyridine)iron (III)-iodide in binary and ternary solvent systems. International Journal of Chemical Kinetics. 2021;**53**(1):16-26

[26] Khattak R, Khan MS, Summer S, Ullah R, Afridi H, Rehman Z, et al. Kinetics of the oxidation of iodide by dicyanobis(phenanthroline)iron(III) in a binary solvent system. International Journal of Chemical Kinetics. 2021;**53**(2): 230-241

[27] Khattak R, Naqvi II, Summer S, Sayed M. Mechanism of the oxidation of 1-(ferrocenyl)-ethanone/ethanol by dicyanobis(phenanthroline)iron(III). Arabian Journal of Chemistry. 2019; **12**(8):4240-4250

[28] Wang X, Stanbury DM. Oxidation of iodide by a series of Fe(III) complexes in acetonitrile. Inorganic Chemistry. 2006; **45**(8):3415-3423

[29] Khattak R, Nazir M, Summer S, Sayed M, Minhaz A, Naqvi II. Thermodynamic aspect: Kinetics of the reduction of dicyanobis(phen) iron(III) by acetylferrocene and

methylferrocenemethanol. Chemical Papers. April 01, 2018;**72**(4):883-893. DOI: 10.1007/s11696-017-0334-1

[30] Domingo PL, Garcia B, Leal JM. Acid–base behaviour of the ferrocyanide ion in perchloric acid media potentiometric and spectrophotometric study. Canadian Journal of Chemistry. 1987;**65**(3):583-589

Chapter 4

The Kinetics of Autoxidation in Wine

Robert E. Coleman, Alexei A. Stuchebrukhov and Roger B. Boulton

Abstract

The kinetics of autoxidation in wine begins with Fenton (1876) who observed that tartaric acid could be oxidized in the presence of iron without peroxide if left in air. Rodopulo (1951) demonstrated that iron tartrate complexes added to wine promoted more extensive oxygen consumption than the molar equivalent of inorganic ferrous or ferric salts. The role of iron complexes in the activation of oxygen, the formation of reactive oxygen species and the initiation of autoxidation are crucial for understanding wine oxidation kinetics. Mechanisms based on hydroxyl radicals versus the ferryl species are likely to have different oxidation products of wine components based on pH effects. The ferryl ion, hydroxyl radical, and tartaric acid radical are proposed as key intermediates in the proposed general mechanism for hydrogen peroxide formation and the autoxidation of wine components. A quantitative kinetic description is presented for the autoxidation of tartaric acid and extended to other acid components as potential ligands. This chapter explores the theoretical considerations of iron complexes formation, oxygen activation, an autoxidative mechanism, and experimental measurements of tartaric acid oxidation as the basis of autoxidation in wine.

Keywords: Fenton, autoxidation, tartaric acid, hydrogen peroxide, radicals, wine

1. Introduction

Wine is an interesting chemical reaction system, in part due to its tartaric acid content. The oxidation of wine is known to be autoxidative [1], stimulated by Fe and Cu ions [2], and is thought to involve Fenton chemistry [3], but neither the rate nor the extent of oxygen consumption can be predicted from a knowledge of pH, metal, phenolic or organic component concentrations. Autoxidation is a spontaneous reaction in air and a radical chain reaction sequence [4, 5].

What is known is that the rate of oxygen consumption in wine is relatively slow in the natural pH range between 3.0 and 4.0. The underlying tartaric oxidation with Fe(II) and oxygen can describe the autocatalytic radical chain reaction sequence, with a distinct initiation stage, a faster, accelerating oxygen consumption propagation stage and a termination stage due to the complete consumption of oxygen and/or Fe(II). It appears that during the propagation stage, ferryl ion, hydroxyl radicals, and/or potentially tartaric acid radicals that are the origins of generating hydroxyethyl radicals when ethanol is present. These hydroxyethyl radicals would lead to an array of selective downstream reactions leading to the collective aged composition of wine. The commonly accepted formation of acetaldehyde would only be one possible fate of this selective radical.

2. Autoxidation and kinetics of tartaric acid

In the study of wine oxidation, the original work by Fenton should be considered as it involves major components found in wine and wine ageing: tartaric acid, iron, and oxygen. It is well known that Fenton used hydrogen peroxide, with Fe(II), to drive the oxidation of tartaric acid. However, he also qualitatively describes the oxidation of ferrous tartrate with air [6], without the addition of hydrogen peroxide, which is now described as autoxidation. It is this reaction and its kinetics with air that sparked further exploration and discussion of pH dependences and autocatalytic kinetics [7, 8]. The study of tartaric acid, iron, and oxygen kinetics under wine-like conditions adds yet another dimension to our understanding of this famous reaction.

Wine-like conditions, in this study [9, 10], restrict the pH to an experimental range from 2.5 to 4.5, while constraining reactants: tartaric acid to 4 g/L (26.7 μM), Fe(II) to 5 mg/L (89.5 μM), and oxygen to ~8.5 mg/L (265 mM). In addition, wine is generally stored in the dark or in darkened bottles to prevent photooxidation. Fenton indicates that the colorimetric response from this fundamental reaction does not appreciably happen without light [11], however with current spectrophotometric instruments, the reaction can be followed without the acceleration from light. The work presented here will also explore a special condition where Fe(II) is equimolar to oxygen, 265 μM, which leads a deeper understanding of the chemical reaction, component limitation and the underlying kinetics.

The time course measurements of dissolved oxygen consumption and Fe(III) formation (**Figure 1**) show the autocatalytic nature of tartaric acid oxidation with three distinct phases: initiation, propagation, and termination. The initiation phase clearly shows a kinetic pH dependence. The work by Smythe [8], proposed the kinetic importance of pH and the Fe(II)-tartrate complex. Tartaric acid, a dicarboxylic acid, exists as three species in this pH range: tartrate (R–), bitartrate (RH-), and tartaric acid (RH_2). With varying pH, the tartrate species concentration changes [12–15], thus changing the free Fe(II), and more importantly Fe(II)-tartrate concentration which correlates with a kinetic role in the initiation. The free Fe(III) and Fe(III)-tartrate ligand(s) concentration also changes across this pH range which increases the intricacy of the reaction mechanics and the intermediate species.

The same time courses (**Figure 1**) show a distinct 1:1 molar relationship between oxygen consumption and Fe(III) formation. It would be expected that Fe(III) formation correlates with Fe(II) consumption. This relationship between iron and oxygen must be maintained when developing a kinetic mechanism for the reaction.

The propagation and termination phases vary with pH as it does with initiation. Increased pH reduces the initiation period, the rate of propagation and the extent to which oxygen is consumed. The termination phase will elucidate additional kinetic considerations when elevated Fe(II) concentrations (265 μM) at pH 2.5 and pH 3.0 (**Figures 2** and **3**) are evaluated. At these levels, the Fe(III) formation terminates as the oxygen concentration is depleted, however beyond this point Fe(III) is consumed, apparently returning to Fe(II). This would indicate a secondary reaction with Fe(III) and an intermediate, thus driving the conversion back to Fe(II).

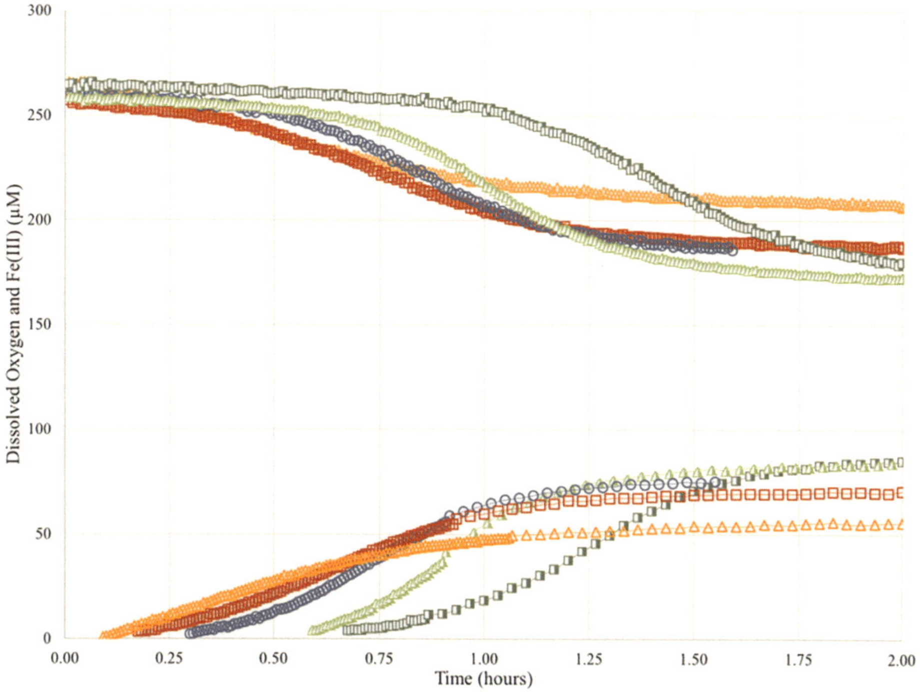

Figure 1.
Oxygen consumption and Fe(III) formation at pH 2.5 to 4.5. Initial conditions at 89.5 μM Fe(II) to initiate the autocatalytic reaction in air-saturated 26.7 mM tartaric acid at pH 2.5 (◨), 3.0 (▲), 3.5 (○), 4.0 (□), 4.5 (△).

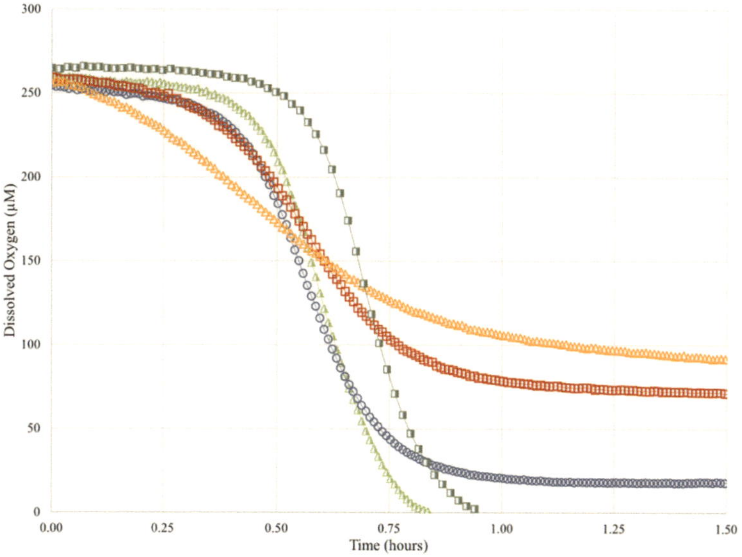

Figure 2.
Oxygen consumption at pH 2.5 to 4.5 with 256 μM Fe(II). Autocatalytic reaction in air-saturated 26.7 mM tartaric acid at pH 2.5 (◨), 3.0 (▲), 3.5 (○), 4.0 (□), 4.5 (△). Reproduced from [10], with the permission of AIP Publishing.

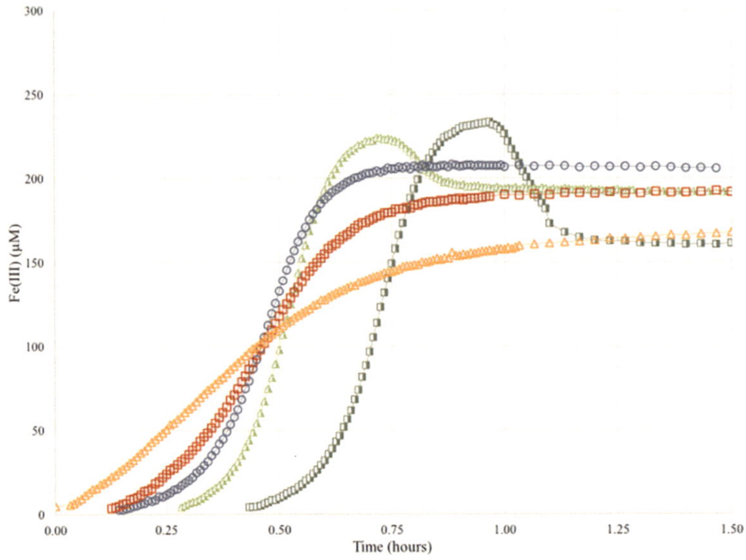

Figure 3.
Fe(III) formation at pH 2.5 to 4.5 with 256 μM Fe(II). Autocatalytic reaction in air-saturated 26.7 mM tartaric acid at pH 2.5 (◨), 3.0 (△), 3.5 (○), 4.0 (□), 4.5 (△). Reproduced from [10], with the permission of AIP Publishing.

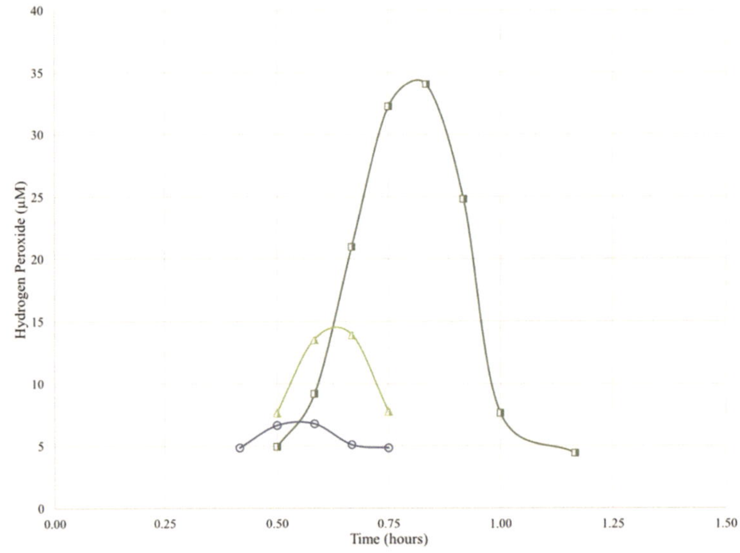

Figure 4.
Hydrogen peroxide formation. Autocatalytic reaction initiated with 265 μM Fe(II) in air-saturated 26.7 mM tartaric acid at pH 2.5 (◨), 3.0 (△), 3.5 (○). Reproduced from [10], with the permission of AIP Publishing.

Hydrogen peroxide has been proposed as a reactant for tartaric acid oxidation [6] and an intermediate in wine oxidation [1, 16]. The simultaneous measurements of hydrogen peroxide at elevated Fe(II) concentrations (265 μM) also show a pH dependency (**Figure 4**). The lower pH levels have a measurable hydrogen peroxide formation with a peak concentration roughly developing as oxygen is depleted. This peak timing can be clearly seen in modeling fitting figures (**Figure 5**). It appears that a

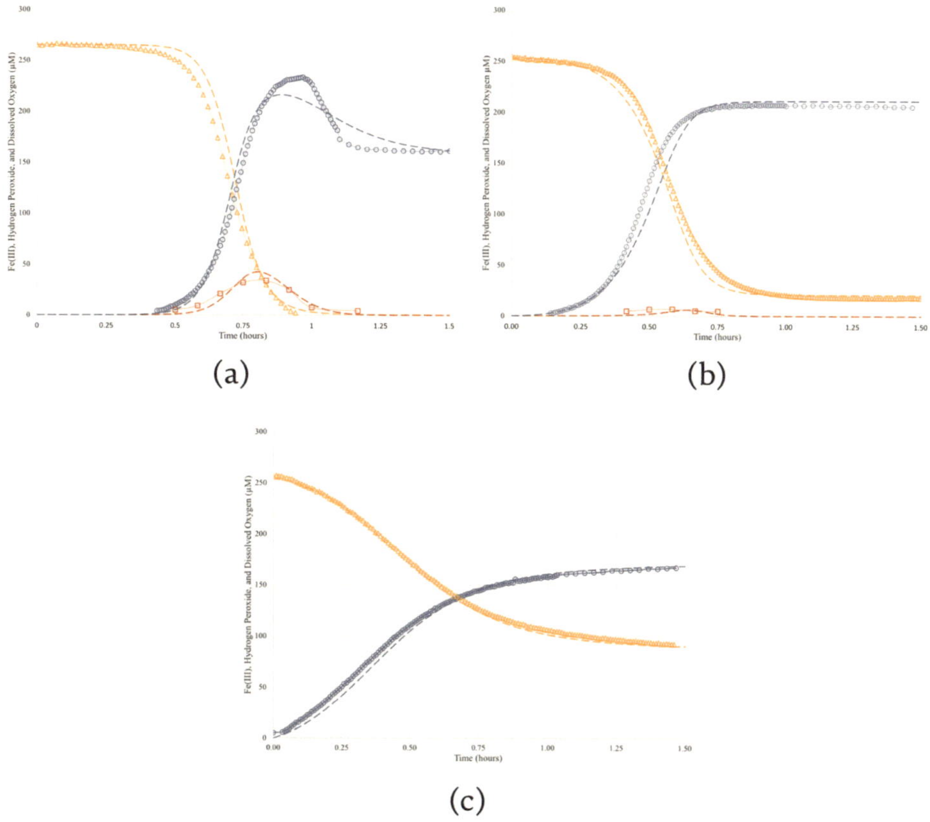

Figure 5.
Fitted consumption and formation curves with proposed mechanism. Dissolved oxygen (△) and predicted (···), Fe(III) (◯) and predicted (···), and hydrogen peroxide (□) and predicted (···) time traces modeled with dissolved oxygen, Fe(III), and hydrogen peroxide time traces at 265 μM initial Fe(II) in air-saturated 26.7 mM tartaric acid at (a) pH 2.5, (b) pH 3.5 and (c) pH 4.5. Reproduced from [10], with the permission of AIP Publishing.

more rapid propagation rate leads to a greater peak hydrogen peroxide concentration. In turn, this peak hydrogen peroxide concentration has a 1:2 molar ratio with the decomposition of Fe(III) (**Figures 3** and **5**). Such a decomposition has been previously explored [17], however oxygen does not appear to be regenerated from the decomposition of hydrogen peroxide in this reaction.

3. Reaction mechanism

Chemical mechanisms for the Fe(II) and oxygen reactions, Fe(II) and hydrogen peroxide reactions, and/or the oxidation of tartaric acid has been explored by several researchers [7, 8, 18–23]. However, these researchers have not had simultaneous measurements of iron, oxygen, and hydrogen peroxide, and the corresponding constraints that come with these time course measurements. These constraints specifically are: pH dependency on all autocatalytic phases, 1:1 molar oxygen consumption to Fe(III) formation during all autocatalytic phases, and 1:2 molar hydrogen peroxide consumption to Fe(III) consumption during termination. The three simultaneous curves and the constraints have led to the proposed comprehensive mechanism in **Figure 6**.

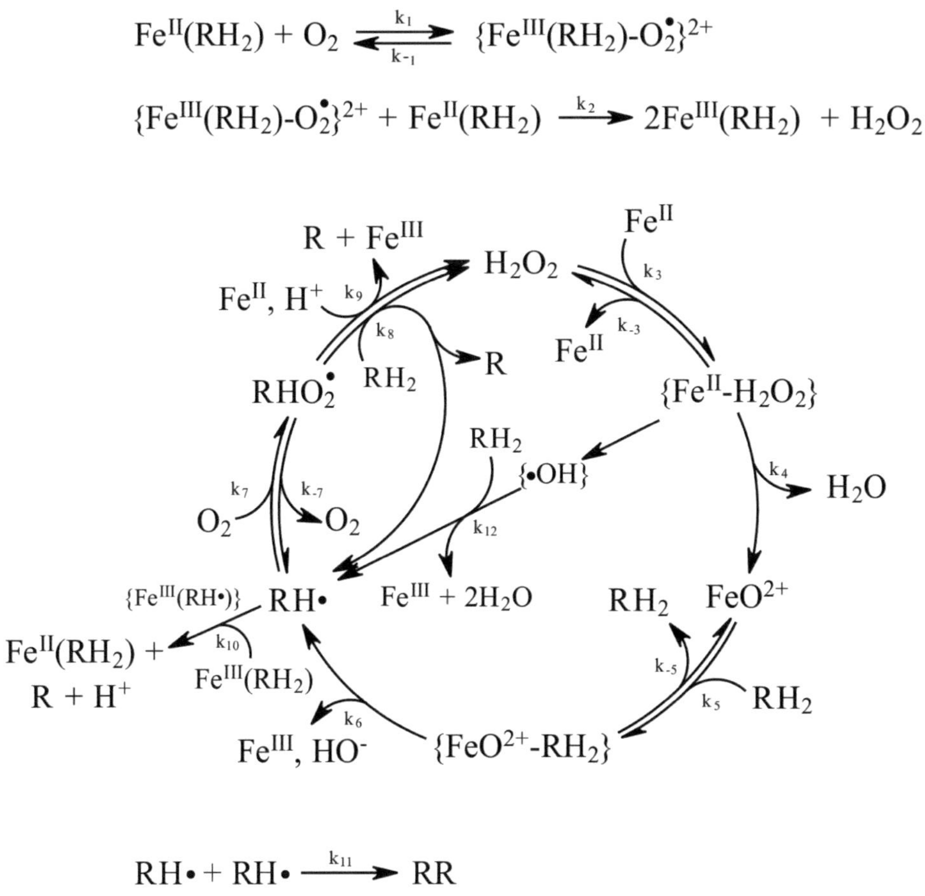

Figure 6.
Mechanism used for the modeling of the oxidation of tartaric acid (RH_2).

The chemical reactions associated with k_1/k_{-1} and k_2 describe the two-electron transfer to oxygen to produce hydrogen peroxide in the initiation phase of the overall tartaric oxidation reaction. These reactions consider the Fe(II) speciation and utilize Fe(II)-tartrate as the oxygen activating species. As the pH increases the concentration of Fe(II)-tartrate increases [12–14], thus leading to a shorter lag time during the initiation phase.

Reactions associated with k_3/k_{-3}, k_4, k_5/k_{-5}, k_6, k_7, k_8, k_9, and k_{12} describe the propagation phase. This scheme describes two alternative propagation pathways; one that utilizes a ferryl (FeO^{2+}) intermediate and another that utilizes a hydroxyl radical (•OH) intermediate. The ferryl intermediate provides an opportunity to explore pH and iron dependency in the propagation phase as it has $pK_a \sim 4.7$ [24]. This pK_a allows for pH varying species concentration in the range of wine pH, whereas the hydroxyl radical will be pH independent. The kinetic modeling work described below will only use the ferryl pathway to fit the Fe(III), oxygen, and hydrogen peroxide simultaneous measurements; the pathway described by hydroxyl radical, k_{12}, will not be explored here.

The oxidation of tartaric acid leads to the formation of dihydroxymaleic acid (R) [6, 11]. The reactions associated with k_8 and k_9 produce dihydroxymaleic acid while continuing to propagate the oxidation cycle by regenerating hydrogen peroxide or regenerating the tartaric radical respectively.

Finally, the reactions described by k_{10} and k_{11} describe termination. The reaction associated with k_{10} terminates the propagation cycle by producing dihydroxymaleic acid, while also regenerating Fe(II) in the process. On the other hand, k_{11} terminates the tartaric radical by allowing for a dimerization to occur (RR). This scheme attempts to provide a mechanism that describes the various constraints provided by the experimental measurements. However, it should be recognized that these reactions may not produce distinct isolatable species as proposed, but reactions may happen within an iron-ligand complex and/or multiple iron-ligand complexes.

3.1 The chain reaction in fenton autoxidation

The stoichiometry indicates that autocatalytic propagation reaction can be described as follows:

$$\mathrm{Fe(II)} + \mathrm{O_2} + \mathrm{H_2O_2} + 2(\mathrm{RH_2}) \rightarrow \mathrm{Fe(III)} + \mathrm{H_2O_2} + 2\mathrm{H_2O} + \mathrm{R} + \mathrm{RH}\bullet \quad (1)$$

(proton needed on the right to balance the charge not shown)

In this reaction, H_2O_2 is regenerated, one electron is taken from Fe(II) and three remaining electrons for O_2 reduction are taken from the substrate: two molecules RH_2 are oxidized with the generation of dihydroxymaleic acid, R, and a tartaric radical, RH•. The generation of a radical will result in the formation of peroxy species RHOO• that would undergo the following oxidation,

$$\mathrm{Fe(II)} + \mathrm{RHO_2\cdot} \rightarrow \mathrm{Fe(III)} + \mathrm{H_2O_2} + \mathrm{R} \quad (2)$$

(again, a proton needed on the left to balance the charge)

which maintains the correct 1:1 oxidation stoichiometry and regenerates H_2O_2. The alternative process with correct stoichiometry is the dimerization of radicals in (Eq. 1) before peroxy radical formation. In addition, other processes such as catalytic reduction of Fe(III) in (Eq. 1) by oxidation of RH•, which would violate 1:1 Fe(II)/O_2 stoichiometry, are possible; however, it appears such processes play only minor role at low pH, as experimental data indicate.

As written, in one cycle of the propagation reaction in (Eqs. 1) and (2), *two* H_2O_2 molecules are generated for each hydrogen peroxide entering the cycle. That means that in one cycle not only H_2O_2 is regenerated, but one *additional* molecule H_2O_2 is formed. This would result in the unlimited exponential growth of hydrogen peroxide in the system, unless some termination/dissipation processes stop the growth. Such processes are chain termination reactions, of which one is the radical dimerization reaction; another is oxidation of RH• by Fe(III) in (Eq. 1). The relative rates of chain multiplication and dissipation/termination define the condition of the exponential growth. The presence or absence of the propagation phase (or its limited form at high pH) observed in the autoxidation reaction can be related to the exponential growth condition in the kinetics using linear stability analysis.

In order to explore the condition of exponential growth we consider a simplified reduced description of the system, keeping track of only most important variables: hydrogen peroxide p1 (h), tartaric radicals p2 (r), oxygen p3 (o), and Fe(III), p4 (f). For hydrogen peroxide and tartaric radicals we have:

$$\begin{aligned} \dot{h} &= V_0 - k_{11}h + k_{12}r \\ \dot{r} &= k_{21}r - k_{22}h - k_{00}r^2 \end{aligned} \quad (3)$$

Here, V_0 is the rate of hydrogen peroxide production by the initiation phase, in the reaction of oxygen and Fe(II); k_{11} is combined rate of conversion of hydrogen peroxide to hydroxyl radical and to ferryl complexes, and also decomposition of hydrogen peroxide by Fe(II); k_{12} is the rate of regeneration of hydrogen peroxide by the reaction of tartaric radicals with oxygen; k_{21} is the rate of generation of tartaric radicals (it may not be exactly same as k_{11}, but close to it); k_{22} is the rate of tartaric radicals removal due to oxidation by Fe(III) (and generation of dihydroxymaleic acid), reaction that competes with oxidation and generation of hydrogen peroxide; and k_{00} is the rate of radical dimerization (these should not be confused with the actual rate constants, k_{11} and k_{12}, in mechanism in the previous section).

As seen, we do not account for all intermediates involved, counting only the initial and final products, partially on the basis that those intermediates are formed on a very short time-scale, such as conversion of hydroxyl radical to tartaric radical, or formation of peroxy-radicals in the reaction of tartaric radicals with oxygen, compared to slow reaction of formation of hydrogen peroxide. The cumulative rates correspond to rate-limiting reactions; for example, formation of hydrogen peroxide from tartaric radical is defined by the proton-coupled electron transfer to peroxy-radical RHOO•. These rates themselves depend on the condition of the reaction, such as pH, and concentration of the substrates; some of them changing significantly in the reaction (oxygen, Fe(II)/Fe(III)), and some do not such as tartaric acid which is in excess. At any given condition, we can assume specific values of these reaction rates and ask what is the kinetics of the system?

The stability of the kinetic system is defined by the linearized equations. Given the current state of the hydrogen peroxide and radicals concentrations, one can ask how a small variation of these concentrations would change the state of the system in time. For small concentrations it is sufficient to consider only a linear part of the system, defined by the kinetic matrix K_{ij} = k_{11}, k_{12}, k_{21}, k_{22}; (or with a modified coefficient k_{22}, $k_{22} + 2k_{00}r$, due to radical dimerization part). The kinetic solution is bi-exponential, the two rates are given by the eigenvalues of the kinetic matrix, and found from the following equation:

$$\begin{aligned} &\det\left(\lambda - \hat{K}\right) = 0 \\ &(\lambda - k_{11})(\lambda - k_{22}) - k_{12}k_{21} = 0 \end{aligned} \tag{4}$$

The populations are changing as combination of two exponentials:

$$p_i(t) = c_{i1}e^{-\lambda_1 t} + c_{i2}e^{-\lambda_2 t} \tag{5}$$

where c_i are some constants. When the product $k_{12}k_{21} = 0$ the two eigenvalues are $\lambda_1 = k_{11}$ and $\lambda_2 = k_{22}$. The two rates describe bi-exponential relaxation of hydrogen peroxide and tartaric radicals to their equilibrium values. However, when $k_{12}k_{21} > 0$, one eigenvalue becomes larger (remaining positive), another becomes smaller and may become negative. In this case the negative eigenvalue gives rise to the exponential growth (and propagation phase of the reaction). The condition for a negative eigenvalue and exponential growth is

$$k_{11}k_{22} < k_{12}k_{21} \tag{6}$$

As we already mentioned, the rate k_{21} is essentially the same as k_{11}, thus the condition is $k_{21} > k_{22}$, that is the rate of generation of radicals is higher than their dissipation. As the determinant of a matrix is a product of its eigenvalues, $\det \hat{K} = \lambda_1\lambda_2$ and one eigenvalue is always positive, the condition of one negative eigenvalue is equivalent to:

$$\det \hat{K} < 0 \tag{7}$$

which is equivalent to condition found in (Eq. 6). The rate coefficients, k_{ij} of the kinetic matrix K, are themselves functions of the conditions of the reaction, which change with time, thus, the above condition may or may not be satisfied at a given every stage of the reaction.

When the exponential growth of hydrogen peroxide and tartaric radicals begins, the dissipation/termination processes get activated and stationary concentrations will be established, until oxygen and Fe(II) diminish. The stationary (maximum) values are found from (Eq. 3) at the stationary conditions $\dot{h} = 0, \dot{r} = 0$; assuming relatively small rate of initiation V_0, the stationary (maximum) values of hydrogen peroxide and tartaric radicals are:

$$\begin{aligned} h_{ss} &\simeq (k_{12}/k_{11})r_{ss} + V_0/k_{11} \\ r_{ss} &\simeq (k_{12} - k_{22})/k_{00} + V_0/(k_{12} - k_{22}) \end{aligned} \tag{8}$$

here we assumed k_{21} to be about the same as k_{11}, and thus in the exponential phase $k_{12} > k_{22}$.

At low pH, the condition of exponential growth appears to be satisfied up to a very low concentration of oxygen; eventually, of course, it breaks down, as k_{12} – rate of regeneration of hydrogen, for which oxygen is needed, diminishes, and rate k_{22} – rate of removal of radicals, bypassing peroxidation, is increasing as Fe(III) increases. The initial lag before fully developed propagation stage is due to the absence of hydrogen peroxide initially, and the incubation period is simply an accumulation of hydrogen peroxide in the system; the exponential multiplication of the initially produced hydrogen peroxide results in the development of the chain reaction that is stabilized by various radical termination process. This fully developed and stabilized chain reaction is what forms the propagation stage of the reaction.

The transition to propagation stage therefore involves a competition between the chain multiplication of radicals and their dissipation. The negative eigenvalue in the kinetic coefficients is a signature of a condition when chain multiplication exceeds that of dissipation.

4. Kinetic modeling

Given the proposed scheme (**Figure 6**), tartaric acid speciation, and iron speciation [12, 13], the experimental Fe(III), oxygen, and hydrogen peroxide simultaneous measurements were fitted using kinetic modeling software, Kintecus [25]. The fitting curves for pH 2.5, 3.5, and 4.5 are shown in **Figure 5** and the resulting kinetic constants are shown in **Table 1**. The modeling provided a reasonable fit for all conditions, especially the highly complex pH 2.5 and 265 μM Fe(II) case where Fe(III) decomposes with a similar timing to hydrogen peroxide.

The ability of the model to fit experimental data across pH provides directional confidence. It should be recognized that the k values are not constant across pH and

k_1/k_{-1}	2.4E+01	M^{-1}
k_2	1.4E+01	$M^{-1} s^{-1}$
k_3/k_{-3}	7.8E+01	M^{-1}
k_4	8.4E+01	s^{-1}
k_5/k_{-5}	6.2E+02	M^{-1}
k_6	4.9E+05	s^{-1}
k_7/ k_{-7}	7.9E+01	M^{-1}
k_8	1.1E+02	$M^{-1} s^{-1}$
k_9	6.6E+03	$M^{-1} s^{-1}$
k_{10}	9.4E+00	$M^{-1} s^{-1}$
k_{11}	1.2E+00	$M^{-1} s^{-1}$

Table 1.
Estimated kinetic constants for 265 μM initial Fe(II) in air-saturated 26.7 mM tartaric acid at pH 2.5.

further examination of species and pH dependent reactions is required, however the robustness of the model and k values within a single pH can be evaluated by making predictions and examining the resulting fit. **Figure 7** shows two predicted curves against experimental measurements where the initial condition of hydrogen peroxide concentration is changed by adding 2.65 μM and 26.5 μM. The predicted and actual measurements are nearly identical, which speaks to the power of the modeling and mechanism.

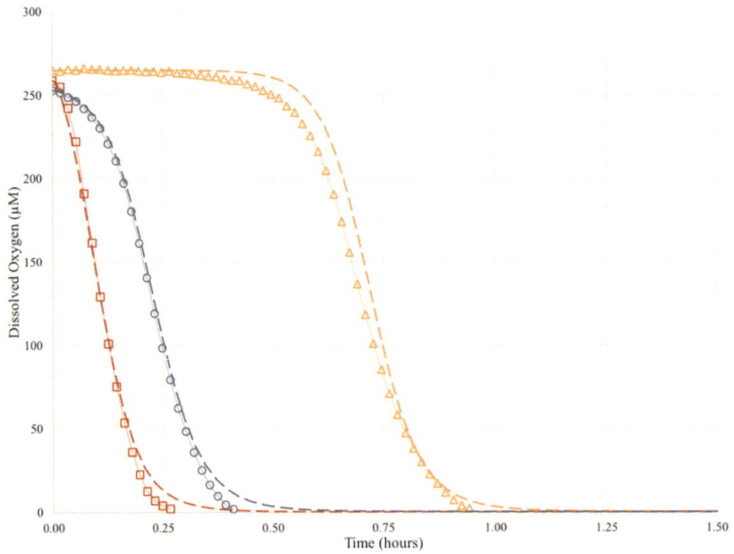

Figure 7.
Oxygen consumption dependence on addition of H_2O_2 and predictions. Time traces with 0 μM (△), 2.65 μM (○), and 26.5 μM (□) hydrogen peroxide added at initiation (t = 0); reaction of 265 μM Fe(II) in air-saturated 26.7 mM tartaric acid at (A) pH 2.5. Fitted trace with 0 μM (···) added hydrogen peroxide. Predicted traces with 2.65 μM (···), and 26.5 μM (···) added hydrogen peroxide. Reproduced from [10], with the permission of AIP Publishing.

DOI: http://dx.doi.org/10.5772/intechopen.103828

5. Discussion

The characteristic sigmoid shape of oxygen consumption in the autoxidation of tartaric acid is observed at all pH conditions. The initiation step is due to oxygen activation to hydrogen peroxide by a Fe(II)-tartrate complex. It is rapid with a time scale of minutes to hours. This reaction step is pH sensitive, slower at lower pH (2.5) than higher (4.5). The extent of peroxide formation is limited to the pool of the Fe(II) state, free and complexed. Initiation is the critical feature of autoxidation, both in these solutions and in wine.

The propagation step is due to the free Fe(II) and hydrogen peroxide oxidation of tartaric acid to produce a tartaric radical which then goes further to form dihydroxymaleic acid, regenerating Fe(II) and consuming additional oxygen as hydrogen peroxide is reformed.

The termination step cannot be explained by the direct decomposition of residual hydrogen peroxide by Fe(III) since oxygen formation is not observed. This termination step may also involve intermediates and end products such as the tartaric acid radical and dihydroxymaleic acid, which adds to the complexity of tartaric oxidation and the accuracy of this model to explain the termination as observed at low pH.

The mechanism proposed here, (**Figure 6**), is slightly different from that previously presented [10]. The difference lies in the propagation steps leading to the exponential growth of one hydrogen peroxide leading to two. The current version does not as accurately fit the loss of Fe(III) observed at the end of oxygen consumption, in the 2.5 pH cases.

The reactivity of dihydroxymaleic acid and the initiation of subsequent radical chain reactions makes describing and interpreting of the termination stage in this system complicated; however, it can be linked to termination of the chain propagation conditions (a negative eigen-value of the kinetic matrix) of our pseudo-first order reaction analysis of Section 3.

5.1 The role of the acid

The unique properties of tartaric acid in these autoxidation reactions can be seen if other major organic acids are substituted for it in the same reaction medium. **Figure 8** shows the individual oxidation of malic, citric and succinic acids at pH 2.5 and 4.5.

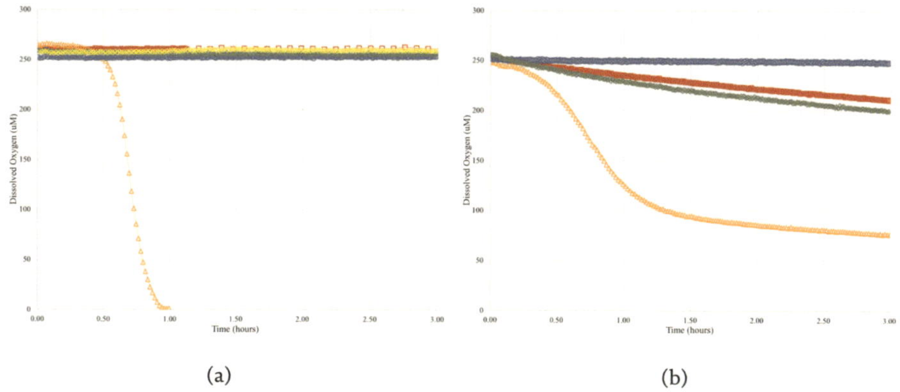

Figure 8.
Wine acids and oxygen consumption. Autocatalytic reaction initiated with 265 μM Fe(II)n air-saturated 26.7 mM tartaric acid (△), malic acid (□), succinic acid (○), citric (◇) and hydrochloric acid (+) at (a) pH 2.5 and (b) pH 4.5. Reproduced from [10], with the permission of AIP Publishing.

At low pH none of these acid show any significant propagation stage while at high pH malic and citric acid show a slow propagation stage but still much slower than tartaric.

The propagation kinetic phase is associated with a chain reaction that requires regeneration of radicals with multiplicity factor greater than one. In our scheme, *two* hydrogen peroxide molecules are regenerated in the chain propagation reaction with tartaric acid (at low pH) for each hydrogen peroxide entering the cycle. Thus multiplicity factor is two. **Figure 9** illustrates the importance of multiplicity factor for a chain reaction. As seen in the figure, upon addition of hydrogen peroxide to citric and succinic acids, the same amount of oxygen is consumed, with ratio 1:1. Here one acid radical is produced per hydrogen peroxide, which reacts further with oxygen to produce peroxy species, but no addition radical or hydrogen peroxide is regenerated. Thus multiplicity factor is zero.

Indeed, the reaction overall oxygen consumption is given by the sum of probabilities of the chain, where the multiplicity factor q (probability of radical generation):

$$\text{Oxygen consumption} = 1 + q + q^2 + q^3 + \cdots = 1/(1 - q) \qquad (9)$$

For malic acid, interestingly, *four* oxygen molecules are consumed for each hydrogen peroxide added, with ratio 1:4. It can be interpreted as a reaction with multiplicity factor q = 3/4, but still less than one, needed to ignite the chain reaction. The overall number of radicals and oxygen consumed is summed to four, as observed. The factor 3/4 is clearly related to the chemical nature of malic acid that has one OH group out of 4 possibilities on C2 and C3 carbons, and 3/4 occupied by hydrogens.

For tartaric acid the multiplicity factor is obviously greater than one and the overall oxygen consumption is as much as 10 in this trial, and the chain could run without stopping until all oxygen or Fe(II) is consumed. That is what we observe indeed for tartaric acid at low pH.

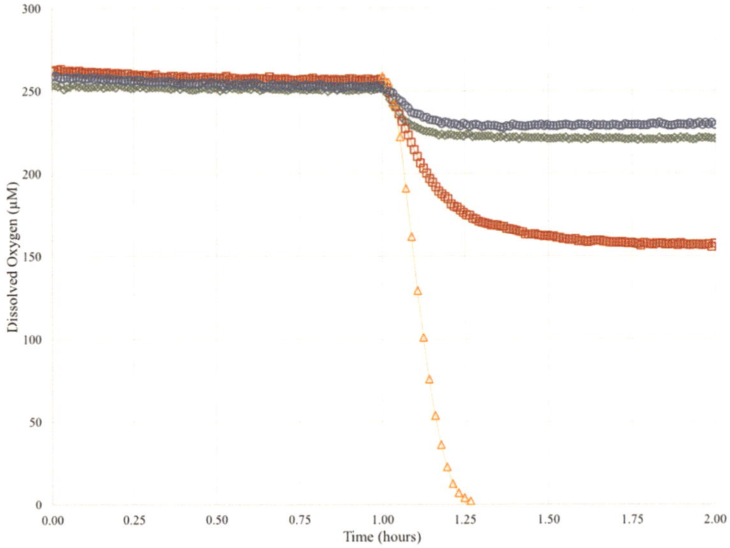

Figure 9.
Wine acids and oxygen consumption initiated with hydrogen peroxide. Autocatalytic reaction initiated with 265 μM Fe(II) and addition of 26.5 μM hydrogen peroxide in air-saturated 26.7 mM tartaric acid (△) offset to 1 hour to align with: malic acid (□), succinic acid (○), and citric acid (◇) at pH 2.5. Reproduced from [10], with the permission of AIP Publishing.

By comparison in case of succinic and citric acids, the overall probability of oxygen consumption equals to 1.

5.2 The role of ethanol

The addition of ethanol to this system (**Figure 10**), shows that even at concentrations of 26.5 mM and both pH of 2.5 and 4.5, the propagation reaction is not established, even when oxygen and Fe(II) are available. This effect, due to ethanol at 100 times the Fe(II) concentration is similar across pH. This can be interpreted as a competition of ferryl ion or hydroxyl radical for ethanol over tartaric acid or the depletion of tartaric radicals in the proposed cycle due to hydroxyethyl radical formation rather than a solvent dielectric effect.

This indicates that the hydrogen peroxide is limiting the propagation stage but the external addition overcomes this limitation in rate (**Figure 11**). The extent of reaction when hydrogen peroxide is available, is independent of pH. In the absence of ethanol,

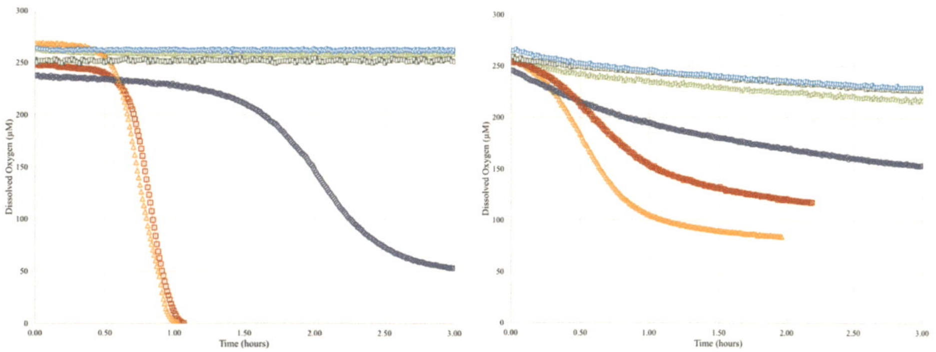

(a) (b)

Figure 10.
Ethanol and tartaric acid. Autocatalytic reaction with 0 mM (△), 0.265 mM (□), 2.65 mM (○), 26.5 mM (▲), 265 mM (◧), and 2.56 M or 15% (v/v) (◑) ethanol with 265 μM Fe(II) in air-saturated 26.7 mM tartaric acid at (a) pH 2.5 and (b) pH 4.5.

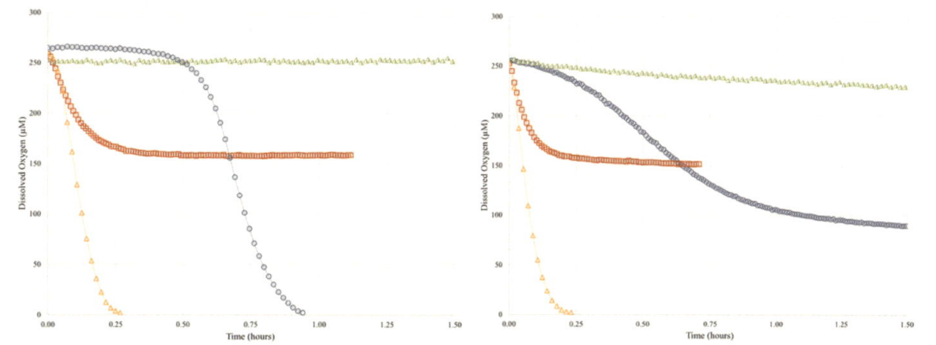

(a) (b)

Figure 11.
Ethanol and tartaric acid initiated with hydrogen peroxide. Autocatalytic reaction with 0 mM ethanol and 0 μM H_2O_2 (○), 26.5 mM ethanol and 0 μM H_2O_2 (▲),0 mM ethanol and 26.5 μM H_2O_2 (△), 26.5 mM ethanol and 26.5 μM H_2O_2 (□) with 265 μM Fe(II) in air-saturated 26.7 mM tartaric acid at (a) pH 2.5 and (b) pH 4.5.

100% of the oxygen would be consumed within 15 minutes and this result is also pH independent.

5.3 The role of sulfur dioxide

The addition of sulfur dioxide at 46.9 μM (30 mg/L) prevents significant development of the propagation stage, at pH 2.5 and 4.5. This is due to the reaction of bisulfite with the hydrogen peroxide formed during initiation, preventing it from accumulating to the level required for a significant propagation reaction to develop. This reaction is known to be rapid [26] at pH 2.5 and 4.5; it is essentially complete within seconds. When the addition is made during the propagation stage, the reaction is terminated even when oxygen and Fe(II) are still present. The implication for wine making is that sulfur dioxide is acting on the hydrogen peroxide production at the point of its formation in the initiation step, essentially preventing the downstream chain reactions that would normally occur (**Figure 12**).

5.4 The role of copper

The addition of copper(II) as $CuSO_4$ displays very different responses at pH 2.5 and 4.5. At pH 2.5 the addition at the beginning of the reaction prevents the development of a propagation stage, suggesting that it is reacting with hydrogen peroxide formed in the initiation stage. This can be overcome by a late addition of hydrogen peroxide wherein the reaction quickly goes to completion – data not shown [9]. A similar result occurs when the addition is in mid-propagation, causing subsequent termination due to its reaction with hydrogen peroxide. In contrast, at pH 4.5 the addition at the beginning results in an enhanced initiation reaction, presumably due to the action of Cu(II)-tartaric complex providing additional oxygen activation and diminished free Cu(II) availability. A late addition of hydrogen peroxide allows the

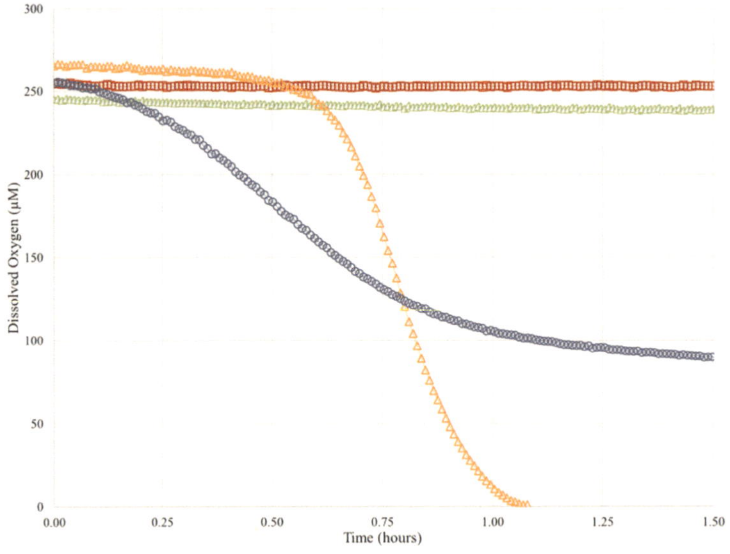

Figure 12.
Oxygen consumption with SO_2 additions. Autocatalytic reaction with 0 μM (△) and 30 mg/L SO_2 (□) at pH 2.5, and 0 μM (○) and 30 mg/L SO_2 (△) at pH 4.5 at initiation (t = 0); reaction of 265 μM Fe(II) in air-saturated 26.7 mM tartaric acid.

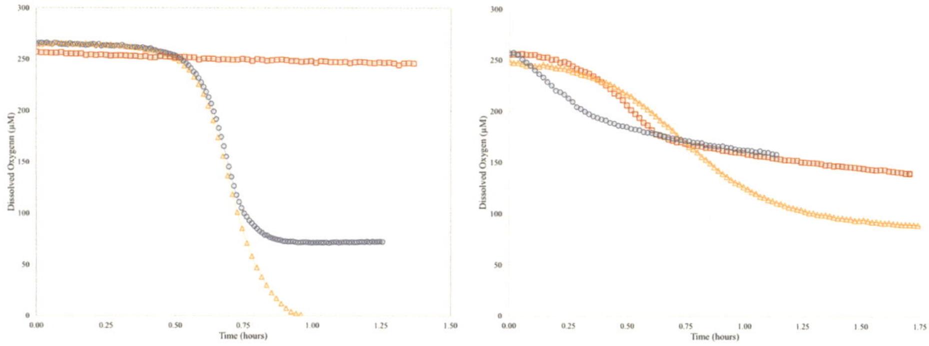

(a) (b)

Figure 13.
Oxygen consumption with copper additions. Time traces with 0 µM (△), 7.87 µM Cu(II)SO_4 (○) at propagation midpoint, and 7.87 µM Cu(II)SO_4 (○) at initiation (t = 0); reaction of 265 µM Fe(II) in air-saturated 26.7 mM tartaric acid at (a) pH 2.5 and (b) pH 4.5.

reaction to go to essentially the same level of completion as when no addition is made [9]. The oxidation of tartaric acid in the presence of Cu(II) is known to occur [27] (**Figure 13**).

5.5 Autoxidation of wine

The connection between the classic tartaric acid oxidation studies of Fenton, Warburg, Wieland and Franke, and Smythe and those occurring in wine comes from investigations of the oxidation of dihydroxymaleic and tartaric acids in the presence of iron (II) salts in wines [28]. Rodopulo [29] described oxidation and rearrangements resulting from tartaric oxidation into intermediates of dioxytartaric, dioxosuccinic, mesoxalic acids and glycolaldehyde and likely final products as glyoxalic and oxalic acids. He noted that in the presence of air dihydroxymaleic spontaneously oxidizes to dioxosuccinic acid, which in the absence of oxygen will react with tartaric acid forming two dihydroxymaleic acids. His opinion was that wines in an anaerobic state would contain dioxosuccinic while those with exposure to air would likely contain traces of glyoxylic and oxalic acids. A key contribution of his work was the demonstration that the addition of a ferrous tartrate precipitate to wine caused further oxygen consumption than that of ferrous sulfate, suggesting the importance of the Fe(II)-tartrate complex in wine oxidation. More accessible descriptions of this finding can be found elsewhere [30, 31]. Baraud [32] further investigated the oxidation of tartaric and dihydroxymaleic acids in wine-like conditions and tried to identify all of the products, including one of the unknown intermediates from these reactions. The relationship between these components has recently been summarized by Duca [27] and referred to as the Baraud cycle.

The proposed (**Figure 6**) incorporates the formation of ferryl ions and radicals, such as the hydroxyl and the tartaric radical, that are expected to be able to extract the α-hydrogen from ethanol to form the 1-hydroxyethyl radical in the presence of even small concentrations of ethanol. Hydroxyethyl radical was identified as the most important radical in beer by Andersen and Skibsted [33] and shown to be the active intermediate in the oxidation of linoleic acid to (E)-2-nonenal, a key impact volatile in "staling" character of beer [34]. It is now known to be the central radical responsible for the selective oxidation of humulones and hop acid components rather than the hydroxyl radical during the oxidation of beer [35].

Hydroxyethyl radical has been shown to react selectively with the flavonols quercetin and kaempferol, but not with epicatechin [36], and while others found that most flavonoids with a C_n-C_{n+1} double bond and caffeic acid were all reactive with the hydroxyethyl radical [37]. The hydroxyethyl radical has been found in a wine held at 55°C, [38] and its selective reaction with some phenol and thiol entities in model wine has been demonstrated [39]. It is known to react with glutathione [40] as well and cinnamates in wine conditions [41]. We propose hydroxyethyl to be the major and more selective oxidizing agent, whose reactivity is likely to be determining the identity and concentrations of the downstream radical products in wine. For this reason we expect there to be little involvement of dihydroxy phenols and tannin in such reactions. As such, hydroxyethyl radical will have a selective influence on specific phenols and glutathione in determining the oxidation outcomes in wine.

The action of sulfur dioxide is to interact with hydrogen peroxide concentrations at the point of initiation and to prevent propagation. The role of ethanol is to divert ferryl ion and/or other radicals into hydroxyethyl radicals and the subsequent radical chain reactions are likely influenced more by reduced glutathione levels and the cinnamates and flavones (but not due to their dihydroxy patterns). As such many of the subsequent reactions associated with oxidation may have little if anything to do with the dissolved oxygen concentration or the quantity of oxygen it has been exposed to.

The autoxidation sequence in wine can be classified into at least 3 periods, those reactions that take place within hours and days, those resulting from that which continue to interact in the days and weeks after, and those that continue to react and rearrange in the subsequent months and years. The first period is the activation of oxygen, the autoxidation described here and the generation of tartaric acid and related radicals. The second period would be the further reactions associated with more stable and long-lived radicals not necessarily in the presence of oxygen and would be selective radical transfer reactions between different wine components, typically not ethanol. The third period would be the long-term aging reactions. These reactions would be disturbed or interspersed with periodic and/or slow diffusional delivery of oxygen, generally in bottles. It is common for wines to be exposed to some oxygen within the winery during transfers, aging and bottling and there can be abrupt increases in concentration of dissolved oxygen or slow diffusional delivery such as though barrels and porous bottle closures.

There appears to be some confusion around the role of certain wine components involved in the initiation reactions compared with those involved in the propagation and termination reactions as well as the time scales involved. It is expected that subsequent radical reactions will continue after the first stage reaction is completed. These will include redox reactions, polymerization reactions and condensation reactions but all would involve relatively stable radicals and are not expected to require additional oxygen. This makes attempts to correlate the extent of product formation in wine with the rate, the extent of oxygen consumption, or the initial wine composition of major components likely to be unsuccessful.

Existing reaction pathways that have been proposed for wine oxidation use mostly free Fe(II) ions as the initial reactive species and all suggest the formation of the hydroxyl radical as the high oxidation state radical [42, 43]. Most of these proposed pathways have a coupled oxidation of a dihydroxy phenol for Fe(II) regeneration but none account for pH or the fact that almost half of the total Fe(II) in wine is in the

Fe(II)-tartrate complex form. None of these proposed pathways can be used to describe the observed autoxidation of tartaric acid in wine conditions. Most propose the formation of acetaldehyde from ethanol as the major oxidation product. The indiscriminate and almost instantaneous reactions with hydroxyl radicals should result in acetaldehyde (and glyceraldehyde) formation directly coupled to oxygen consumption and in a ratio of products proportional to their initial concentrations. Formation of acetaldehyde involving dihydroxy phenols does occur but only at elevated temperatures, 50°C [1]. Such formation has not been shown to be related to either the extent or rate of oxygen consumption at ambient or wine storage temperatures.

The role of other transition metal complexes in the initiation and propagation reactions needs to be investigated further. Here, there are effects due to the presence of malic acid and copper (II) which will vary between wines, especially before and after the malolactic fermentation and as a result of copper additions during winemaking. We expect their effects to be related to the concentrations of their complexes.

Lastly, the recovery of wines to an initial state after exposure to oxygen was reported to take several days [16]. This might be interpreted as being due to a slower return of Fe(II) for further initiation and or propagation reactions due to certain wine constituents that are absent in these model solutions studies. This deserves further research attention.

6. Conclusions

The autoxidation of tartaric acid in the presence of Fe(II) has been demonstrated in solution over the pH range 2.5 to 4.5. A mechanism that can describe the observations is proposed and fitted to a kinetic model. Preliminary estimates of the rate constants are presented. The effects on these reactions due to additions of ethanol, sulfur dioxide and copper at wine-like conditions are described. The implications of this radical chain reactions sequence to describe the autoxidation of wine and idea of radical chain propagation in wine are discussed.

Acknowledgements

This work was supported in part by Treasury Wine Estates (REC), by NIH grant GM054052 (AAS) and the Stephen Sinclair Scott Endowment (RBB).

Conflict of interest

The authors declare no conflict of interest.

Author details

Robert E. Coleman[1], Alexei A. Stuchebrukhov[2] and Roger B. Boulton[1*]

1 Department of Viticulture and Enology, University of California, Davis, California, CA, USA

2 Department of Chemistry, University of California, Davis, Davis, CA, USA

*Address all correspondence to: rbboulton@ucdavis.edu

References

[1] Wildenradt HL, Singleton VL. The production of aldehydes as a result of oxidation of polyphenolic compounds and its relation to wine aging. American Journal of Enology and Viticulture. 1974; **25**(2):119-126

[2] Warburg O. Über die Rolle des Eisens in der Atmung des Seeigeleis nebst Bemerkungen über einige durch Eisen beschleunigte Oxydationen. In: Über die Katalytischen Wirkungen der Lebendigen Substanz. Berlin, Heidelberg: Springer; 1928. pp. 47-66. DOI: 10.1007/978-3-642-47774-4_4

[3] Haber F, Weiss J. Über die Katalyse des Hydroperoxydes. Die Naturwissenschaften. 1932;**20**(51): 948-950. DOI: 10.1007/BF01504715

[4] Simic MG. KM. Autoxidation in Food and Biological Systems. Boston, MA: Springer; 1980. DOI: 10.1007/978-1-4757

[5] Simic MG. TKA. Oxygen Radicals in Biology and Medicine. Boston, MA: Springer; 1988. DOI: 10.1007/978-1-4684-5568-7

[6] Fenton HJH. Oxidation of tartaric acid in presence of iron. Journal of the Chemical Society. 1894;**65**:899-910. DOI: 10.1039/ct8946500899

[7] Wieland H, Franke W, den Mechanismus Ü, der Oxydationsvorgänge. XIV. Die Activierung des Sauerstoffs durch Eisen. Liebigs Annalen der Chemie. 1928;**464**:101-226. DOI: 10.1002/ jlac.19284640109

[8] Smythe CV. The mechanism of iron catalysis in certain oxidations. The Journal of Biological Chemistry. 1931; **90**(1):251-265

[9] Coleman R. Kinetics of Oxygen Consumption in Solutions of Iron and Tartaric Acid. Davis: University of California; 2019

[10] Coleman RE, Boulton RB, Stuchebrukhov AA. Kinetics of autoxidation of tartaric acid in presence of iron. The Journal of Chemical Physics. 2020;**153**(6). DOI: 10.1063/5.0013727

[11] Fenton HJH. The constitution of a new dibasic acid, resulting from the oxidation of tartaric acid. Journal of the Chemical Society. 1896;**69**:546-562. DOI: 10.1039/ct8966900546

[12] Timberlake CF. Iron–tartrate complexes. Journal of the Chemical Society. 1964:1229-1240. DOI: 10.1039/ jr9640001229

[13] Yokoi H, Mitani T, Mori Y, Kawata S. Complex formation between iron(III) and tartaric and citric acids in a wide pH range 1 to 13 as studied by magnetic susceptibility measurements. Chemistry Letters. 1994;**23**(2):281-284. DOI: 10.1246/cl.1994.281

[14] Green RW, Parkins GM. Complexes of iron with d-tartaric and meso-tartaric acids. The Journal of Physical Chemistry. 1961;**65**(9):1658-1659. DOI: 10.1021/ j100905a523

[15] Salvadó V, Ribas X, Blanco M, Valiente M. On the chemistry of iron in biosystems. I. Complex formation between Fe(III) and tartaric acid: a 'core + link' mechanism. Inorganica Chimica Acta. 1987;**137**(3):155-159. DOI: 10.1016/ S0020-1693(00)81159-7

[16] Ribéreau-Gayon J. Contribution à l'étude des oxydations et réductions dans les vins (Thèse Doctorat ès Sciences Physiques). Bordeaux, France:

Bordeaux University Institute of Enology; 1933

[17] Bohnson VL, Robertson AC. The catalytic decomposition of hydrogen peroxide by ferric salts II. Journal of the American Chemical Society. 1923; **45**(11):2493-2503. DOI: 10.1021/ja01664a003

[18] Haber F, Weiss J. The catalytic decomposition of hydrogen peroxide by iron salts. Proceedings of the Royal Society London A. 1934;**147**(861): 332-351

[19] Liu X. Progress in the mechanism and kinetics of Fenton reaction. MOJ Eco. Environmental Sciences. 2018;**3**(1). DOI: 10.15406/mojes.2018.03.00060

[20] Wells CF. The mechanism of the autoxidation of ferrous ions in aqueous solution. Journal of Inorganic and Nuclear Chemistry. 1968;**30**(3):893-894. DOI: 10.1016/0022-1902(68)80455-5

[21] George P. A comparison of the decomposition of hydrogen peroxide by catalase, ferrous and ferric ions, haemin and ferrous phthalocyanine. The Biochemical Journal. 1948;**43**(2): 287-295. DOI: 10.1042/bj0430287

[22] Clark A, Prenzler P, Scollary G. Impact of the condition of storage of tartaric acid solutions on the production and stability of glyoxylic acid. Food Chemistry. 2007;**102**(3):905-916. DOI: 10.1016/j.foodchem.2006.06.029

[23] Koppenol WH. The centennial of the Fenton reaction. Free Radical Biology & Medicine. 1993;**15**(6):645-651. DOI: 10.1016/0891-5849(93)90168-t

[24] Silaghi-Dumitrescu R, Reeder Brandon J, Nicholls P, Cooper Chris E, Wilson MT. Ferryl haem protonation gates peroxidatic reactivity in globins. The Biochemical Journal. 2007;**403**(3): 391-395. DOI: 10.1042/bj20061421

[25] Ianni JC. Kintecus, Version 6.7. Available from: www.kintecus.com. 2018

[26] McArdle JV, Hoffmann MR. Kinetics and mechanism of the oxidation of aquated sulfur dioxide by hydrogen peroxide at low pH. The Journal of Physical Chemistry. 1983;**87**(26): 5425-5429. DOI: 10.1021/j150644a024

[27] Duca G. Homogeneous Catalysis with Metal Complexes in the Chemical Industry and Foodstuffs Chemistry. In: Springer Series in Chemical Physics. Vol. 102. Berlin, Heidelberg: Springer; 2012. DOI: 10.1007/978-3-642-24629-6_7

[28] Ribereau-Gayon J. Trait d'Oenologie, Transformations et Traitements des Vins. Paris: Librarie Polytechnique Ch. Beranger; 1950

[29] Rodopulo AK. Oxidation of tartaric acid in wine in the presence of heavy metal salts (activation of oxygen by iron). Journal: Izvestiya Akademii Nauk SSSR. 1951;**3**:115-128

[30] Farkaš J. Technology and Biochemistry of Wine. New York: New York: Gordon and Breach Science Publishers; 1988

[31] Amerine MA. In: Amerine MA, Joslyn MA, Joslyn MA, editors. Table wines; the technology of their production [by]. Berkeley: University of California Press; 1970

[32] Baraud J. Étude des dérivés naturels de l'acide tartrique. Annali di Chimica. 1954;**9**:535-604

[33] Andersen ML, Skibsted LH. Electron spin resonance spin trapping identification of radicals formed during

aerobic forced aging of beer. Journal of Agricultural and Food Chemistry. 1998; **46**(4):1272-1275. DOI: 10.1021/jf9708608

[34] Vanderhaegen B, Neven H, Verachtert H, Derdelinckx G. The chemistry of beer aging–a critical review. Food Chemistry. 2006;**95**(3):357-381. DOI: 10.1016/j.foodchem.2005.01.006

[35] de Almeida NE, Homem-de-Mello P, De Keukeleire D, Cardoso DR. Reactivity of beer bitter acids toward the 1-hydroxyethyl radical as probed by spin-trapping electron paramagnetic resonance (EPR) and electrospray ionization-tandem mass spectrometry (ESI-MS/MS). Journal of Agricultural and Food Chemistry. 2011;**59**(8): 4183-4191. DOI: 10.1021/jf104797d

[36] Marfak A, Trouillas P, Allais DP, Calliste CA, Cook-Moreau J, Duroux JL. Reactivity of flavonoids with 1-hydroxyethyl radical: A γ-radiolysis study. Biochimica et Biophysica Acta. 2004;**1670**:28-39. DOI: 10.1016/j.bbagen.2003.10.010

[37] Hryntsevich I, Shadyro O. Reactions of α-hydroxyethyl radicals with flavonoids of various structures. Bioorganic & Medicinal Chemistry Letters. 2005;**15**:4252-4255. DOI: 10.1016/j.bmcl.2005.06.065

[38] Elias RJ, Andersen ML, Skibsted LH, Waterhouse AL. Key factors affecting radical formation in wine studied by spin trapping and EPR spectroscopy. American Journal of Enology and Viticulture. 2009;**60**(4):471-476

[39] Kreitman GY, Laurie VF, Elias RJ. Investigation of ethyl radical quenching by phenolics and thiols in model wine. Journal of Agricultural and Food Chemistry. 2013;**61**(3):685-692. DOI: 10.1021/jf303880g

[40] Stoyanovsky DA, Wu D, Cederbaum AI. Interaction of 1-hydroxyethyl radical with glutathione, ascorbic acid and α-tocopherol. Free Radical Biology & Medicine. 1998;**24**(1): 132-138. DOI: 10.1016/s0891-5849(97)00205-0

[41] Gislason NE, Currie BL, Waterhouse AL. Novel antioxidant reactions of cinnamates in wine. Journal of Agricultural and Food Chemistry. 2011;**59**(11):6221-6226. DOI: 10.1021/jf200115y

[42] Danilewicz JC. Review of reaction mechanisms of oxygen and proposed intermediate reduction products in wine: Central role of iron and copper. American Journal of Enology and Viticulture. 2003;**54**(2):73-85

[43] Elias RJ, Waterhouse AL. Controlling the Fenton reaction in wine. Journal of Agricultural and Food Chemistry. 2010;**58**(3):1699-1707. DOI: 10.1021/jf903127r

Chapter 5

The ESIPT-Steered Molecular Chameleon for Cations and Anions Based on Alizarin and Alizarin-S: A Comparative Study

Veikko Uahengo

Abstract

Chemosensing properties of Alizarin (**A3**) and Alizarin S (**AS3**) towards anions and cations in acetonitrile are reported. The absorption and fluorescence properties of the two molecular entities were investigated in CH_3CN. Based on the excited state intermolecular proton transfer system (ESIPT), the probes were able to collectively discriminate specific cations and anions via colorimetric observations and spectrometric activities. The investigation revealed that **A3** was selective to Cu^{2+}, Fe^{3+}, and Fe^{2+}, compared to Cu^{2+}, Zn^{2+}, Fe^{3+}, and Ni^{2+} for **AS3.** The disagreement in spectral responses were ascribed to the strong electron withdrawing group present in **AS3**, hence the difference in behaviors. Moreover, the emission properties displayed by **A3** and **AS3** upon molar titrations with cations, were closely similar for all cations, which all nearly experienced fluorescence quenching, except for Zn^{2+} with **A3**, which exhibited fluorescence enhancement. Similarly, a two-step fluorescence effect was observed in **A3** towards anions, which experienced both fluorescence quenching and enhancement, with incremental additions. The simultaneous fluorescence effects were ascribed to the deprotonation activities experienced by **A3**, as excess anion quantities were added. Thus, the sulfonyl electron withdrawing group had an effect on the Alizarin structure, towards the discrimination of anions and cations, both colorimetrically and fluorometrically.

Keywords: ESIPT process, alizarin probe, Fluorogenic sensor, molecular chameleon

1. Introduction

Alizarin is a stable organic compound, prominently known as a red dye with significant industrial applications, particularly its use in dying textile fabrics. The application in textile coloration industry is inspired by the fact that alizarin is a natural compound often referred a natural dye, initially extracted from the roots of plants of the madder genus [1], before it was synthetically made [2], thus, it is a molecular species from nature, exhibiting and portraying green chemistry properties. The molecular structural framework of alizarin is characterized by the anthraquinone moiety bearing two para-positioned intermolecular hydroxyl groups both on one carbocyclic ring adjacent to the quinone ring. A typical excited state intermolecular proton transfer system (ESIPT), alizarin is a natural dye which has been widely used in pigments, as anticancer agents as well as chemical reagents for use in data recording and storage materials due to its tunable electronic properties [3, 4]. In addition, the natural dye has strong antigenotoxic activity, ascribed to the transfer of ultrafast electrons to TiO_2-based materials, which can also perfectly fit as an excellent photosensitizer in dye-sensitized solar cells. Thus, alizarin chromophore has been favored by many researchers, both experimentally and theoretically [5–10].

Ideally, alizarin forms an intramolecular hydrogen bond between a hydroxyl and a carbonyl group, in the ground and excited states, whereby upon photoexcitation, a proton transfer from the hydroxyl to the carbonyl group is observed, which normally results in dual emission bands of the locally excited (LE) and proton-transferred (PT) tautomers [9, 11–13]. Characteristically, this process in known as ESIPT, which is viewed as a very fast photo-tautomerization process taking place along an intramolecular hydrogen bond between two atoms that are significantly tuned by electronic excitation. In recent studies, the practical and applications of the ESIPT mechanism based on their photophysical characteristics and properties have been extensively explored and investigated, especially in laser dyes, OLEDs, molecular switches, fluorescence sensors, and particularly biological systems [14–20]. More importantly, the ESIPT based reactions increase the acidity of the proton donor groups, due to the change of electron density after electron excitation, and the basicity of the acceptor groups is significantly increased to promote the formation of tautomer by intramolecular proton transfer [4, 21–26].

On the other hand, molecular recognition has been the epic center of supramolecular chemistry due to its significant role in biological and environmental systems, through the host-guest interaction chemistry. Consequently, chemosensors are designed for specific target analytes based on their chemical make-up and complementarity towards each other. The impact of sensing biologically important anions such as acetate, cyanide, fluoride, dihydrogen phosphate, etc., have been receiving attention in literature and many industrial applications. A large volume of colorimetric and fluorometric probes for anions such as fluoride (F-), cyanide (CN-), acetate (AcO-), dihydrogen phosphate ($H_2PO_4^-$), hydroxide (OH-) and others have been developed. Hydroxide ions play a very significant role in environmental and physiological systems, thus monitoring its concentration in these systems must be highly prioritized [27–37]. Moreover, the presence of soft (donor) atoms such as oxygen from hydroxyl and quinone groups of the carboxylic ring raises the prospect of dual sensing, for both cations and anions, which stems from the presence of both, the anion receptors (-OH) and the cationphilic groups, through coordination induced interaction [38–41].

Herein, we have conducted a comparative study for the two alizarin-based derivatives, **A3** and **AS3** (**Figure 1**), to investigate their chemosensing properties. The study literally focuses on the effect of the sulfonyl hydroxide group (-SO_3H) present in **AS3**, which is the only difference between the two chemical entities. The two entities are highly stable and available commercially, which are very rich in hydroxyl groups, paving ways for possible hydrogen bonding based charge transfer bonding. The two dyes displayed interesting behaviors in the presence of anions and cations, in water-soluble acetonitrile (CH_3CN) solvent, with certain commonalities and variations upon interacting. Thus, the two probes (**A3** and **AS3**) can be used as colorimetric/fluorometric probes for discriminating specific cations and anions, with distinctive color changes in organic-aqueous solvents.

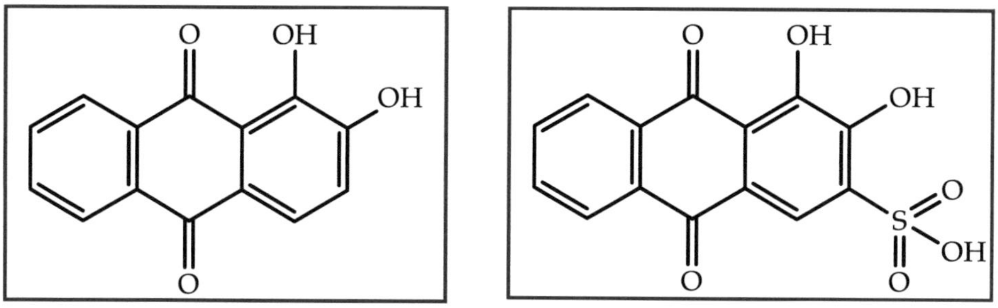

A3 AS3

(a) (b)

Figure 1.
*The molecular 2-D structures of (a) alizarin (**A3**) and (b) alizarin S (**AS3**).*

2. Experimental section

2.1 Materials and apparatus

Compound **A3** and **AS3** were obtained from a commercial source. All the chemicals and reagents used in this study were all of analytical reagent grade. The anions of Cl^-, CN^-, OH^-, AcO^-, Br^-, I^-, $H_2PO_4^-$, HSO_4^-, N_3^-, NO_3^-, OCN^- and F^- were purchased as tetrabutylammonium salts from Sigma–Aldrich. Absorption measurements were performed on Perkin Elmer Lambda 35 spectrophotometer in a standard 3.0 ml quartz cuvette with 1 cm path length at room temperature. Fluorescence measurements were carried out on a steady state excitation and emission spectra on the Molecular Device SpectraMax M2, Plate Reader.

2.2 Procedures for UV: Vis experiments

All UV–Vis spectra were recorded in acetonitrile (CH_3CN) solvent on a Perkin Elmer Lambda 35 spectrometer by adding Tetrabutylammonium salts while keeping the concentration of **AS** or **AS3** constant (1×10^{-5} M). Tetrabutylammonium salt (TBA) anions of Cl^-, CN^-, OH^-, AcO^-, Br^-, I^-, $H_2PO_4^-$, HSO_4^-, N_3^-, NO_3^-, OCN^- and F^- were used for UV–Vis experiments. In addition, a range of heavy metal cation salts ($AgNO_3$, $Al(NO_3)_3$, $Co(NO_3)_2$, $Cr(NO_3)_3$, $Cu(NO_3)_2$, $Fe(NO_3)_2$, $Hg(NO_3)_2$, $MnCl_2$, $Ni(NO_3)_2$, $Pb(NO_3)_2$, $SnCl_2$, $Zn(NO_3)_2$, $CdCl_2$) were used for UV–Vis titrations.

3. Results and discussions

3.1 Photophysical studies of A3 and AS3 with anions and cations

*3.1.1 Visual observation of **A3** and **AS3** with anions and cations*

In order to establish the occurrence of chemical interactions between **A3** and the anions, a series of prepared anionic and cationic solutions (0.03 M) in CH_3CN were tested separately, about 3 ml of **A3** (1×10^{-3} M) in CH_3CN, at room temperature. The colorimetric activities observed were recorded (**Figure 2**). The addition of anions

(a)

(b)

Figure 2.
Observable colorimetric changes of different anions upon interacting with (a) **A3** *and (b)* **AS3** *both (1×10^{-5} M) in CH_3CN at room temperature.*

(Cl^-, CN^-, OH^-, AcO^-, Br^-, I^-, $H_2PO_4^-$, HSO_4^-, NO_3^-, ClO_4^-, N_3^-, and F^-) to **A3**, were investigated through naked eye observable color changes. The dropwise addition of anions (TBA salts) to the shinny-yellow colored **A3** solution, resulted in a series of naked eye observable color changes, ranging from deep blue (CN^-), blue-violet (AcO^-), blue-brownish ($H_2PO_4^-$), brown-yellowish (N_3^-), deep blue (OH^-) and (blue-violet (F^-) as displayed (**Figure 2**). The intensity of the colors is an indication that **A3** is a color-based indicator, with such unique concentrated color change. Clearly, the color changes were due to chemically associated interaction between **A3** and the anions. However, none of the other anions (Cl^-, Br^-, HSO_4^-, NO_3^- and ClO_4^-) used could induce any significant color changes, even when large quantities were added. Interestingly, **A3** was discriminatingly selective and sensitive to N_3^-, turning the yellow color **A3** to dark-yellowish **A3-N3** complex. Probes sensitive and selective to N_3^- are extremely rare in literature, with this being the least in literature reports. In addition, the addition of OCN^- to **A3** displayed a light brownish color of the complex **A3-OCN**. The two anions formed distinctive colors on interaction with **A3**, defying the trend with other anions above, which formed strong blue to violet color upon complexed (**Figure 2a**).

Moreover, comparative colorimetric studies were conducted for **AS3** and anions (**Figure 2b**), were visual observation clearly showed the variation between the two-alizarin derivatives (**A3** and **AS3**). Like with **A3**, the addition of anions to yellow **AS3** resulted in a variety of colorimetric observations, such as F^- (deep purple), AcO^- (pink), $H_2PO_4^-$ (dark yellow), N_3^- (reddish-brown), OCN^- (reddish-brown), OH^- (deep-blue), CN^- (purple) and Br^- (no change) as displayed (**Figure 2b**). The effect of the sulfonyl group on the alizarin molecule was clearly visible upon interacting with anions, giving different colors between **A3** and **AS3**. Apart from the OH^- and F^- ions, which displayed deep-blue and violet colors for both **A3** and **AS3** respectively, the rest of the anionic interactions displayed different colors. However, the pattern in colors are still displayed among the two probes, but clearly the intensities differed perhaps due to different association constants, where some are more strongly associated comparing to others.

Furthermore, the multi-colorimetric sensor (**A3**) was also able to selectively and sensitively detect the presence of cations in acetonitrile. The gradual addition of heavy metal cations, of divalent in nature, to the **A3** solution saw the yellow color changes to varieties of colors ranging from light-green (Cu^{2+}), deep yellow (Zn^{2+}), dark brown (Fe^{3+}) and light orange (Ni^{2+}) as displayed (**Figure 3a**). The range of different colors could mean diverse interacting modes with the probe, or dissimilar geometrical complementarities towards each cation. In addition, the presence of the sulfonic acid group in **AS3** could induce mole or less similar cation interactions, even though different colors, with Cu^{2+} (light green), Zn^{2+} (crimson red), Fe^{3+} (brown) and Ni^{2+} (light orange). In addition, there was a noticeable additional color activities when Hg^{2+} was added to **AS3**, the color from light yellow (**AS3**) to slight deep yellow of the complex formed (**AS3-Hg**) as displayed (**Figure 3b**). The variations in colors of the complexes formed upon the two probes interacting with cations, signifies their differences in coordination and the effect of the sulfonic acid group present in **AS3**.

3.1.2 Absorption properties of **A3** *and* **AS3**

Spectrally, the two probes were characterized by more or less similar absorption spectra, both of them defined by the $\pi \rightarrow \pi^*$ transitions in the ultraviolet region, as well as the internal charge transfer band (ICT) in the visible region. Specifically,

(a)

(b)

Figure 3.
Observable colorimetric changes of different cations upon interacting with (a) **A3** *and (b)* **AS3** *both* (1×10^{-5} *M) in* CH_3CN *at room temperature.*

AS3 displayed a high-energy intense peak centered at 249 nm and another moderate absorption band at 427 nm (**Figure 4a**), with a visual yellow color of the probe, (**Figure 4a** inset). Similarly, **A3** was characterized by an intense $\pi \rightarrow \pi^*$ band at 246 nm, as well the internal charge transfer band at 420 nm (**Figure 4b**), with the color displayed (**Figure 4b**). In addition, in both cases (**A3** and **AS3**) each probe has a light hump at 328 nm and 330 nm respectively, ascribed to the admixture of the $\pi \rightarrow \pi^*$ and the ICT transition overlaps. The only difference in the two spectra is the slight red shift of the spectra of **AS3** with about 2 nm of all the bands.

3.1.3 Absorption studies of **A3** *and* **AS3** *on interaction with anions*

The interaction of **A3** with anions was characterized by similar changes in all four titrations. For instance, the molar addition of 5 equiv. of CN^- to **A3**, resulted in the gradual disappearance of the ICT band at 420 nm, concomitant with the appearance of a band associated with 328 nm peak centered now at 350 nm, as well as a completely new band deep in the visible light region centered at 570 nm (**Figure 5a**). In addition, an intense $\pi \rightarrow \pi^*$ band at 246 nm disappeared followed by another new peak at 260 nm, upon the molar addition of CN^-. Evidently, several isosbestic points were observed at 222 nm, 254 nm, 292 nm, 390 nm and 466 nm, which testimony to the co-existence and formation of new complexes, at equilibrium (**Figure 5a**). Similar patterns were observed upon gradual addition of AcO^-, F^- and OH^- (**Figure 5b–d**), which were all due to the hydrogen bonding interaction of these anions with the hydroxyl groups of **A3**. It is evident that the interactions were similar in nature, given the similarities of the absorption spectra in all four cases, which are all via hydrogen bonding and suspectedly deprotonations.

On the other hand, the molar titrations of **AS3** with anions resulted in similar patterns as those of **A3** for the three anions (F^-, CN^-, OH^-) as displayed (**Figure 6a, c,** and **d**). The molar titration of **AS3** with F^- saw a gradual disappearance of the ICT peak at 427 nm, concomitantly with the appearance of new peaks at 350 nm and 550 nm respectively (**Figure 6c**). The spectral activities are ascribed to the hydrogen bonding induced charge transfer upon bonding has taken place, to form complex pedants of **AS3-F**, **AS3-CN** and **AS3-OH**. Like in **A3**, the spectral activities were accompanied by several isosbestic points,

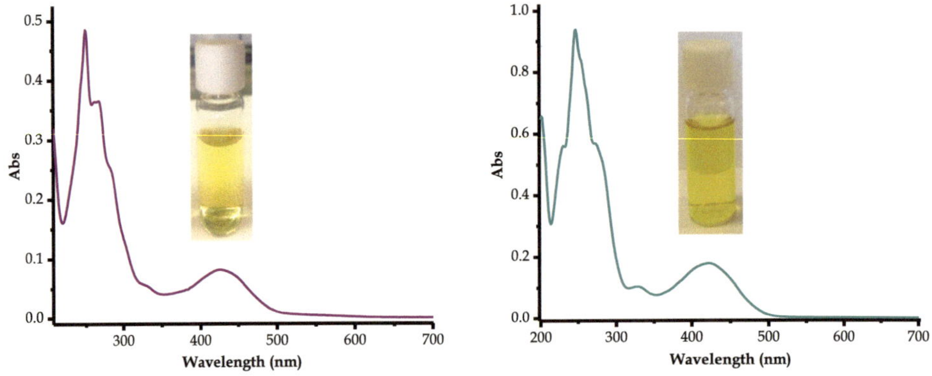

(a) (b)

Figure 4.
Absorption spectra of (a) **AS3** *(1×10^{-5} M) and (b)* **A3** *(1×10^{-5} M), both in CH_3CN. The insets are colorimetric signatures under daylight conditions.*

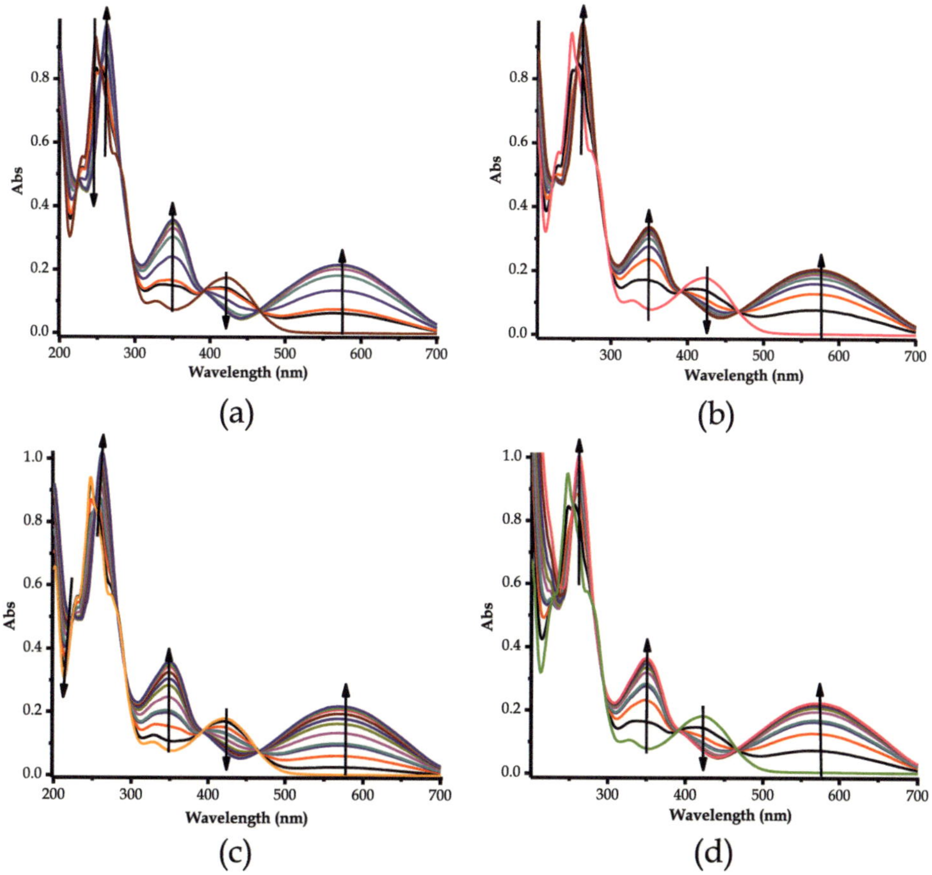

Figure 5.
The absorption titration spectra of **A3** *(1 x 10^{-5} M) in CH_3CN, with 5 equiv. of (a) CN^-, (b) AcO^-, (c) F^- and (d) OH^- at room temperature.*

at 238 nm, 300 nm, 400 nm and 463 nm, which proves the formation of new pedants co-existing with other species at equilibrium. The other two titrations have resulted in similar patterns like that of F^- (**AS3-F**), with spectral shifts precisely resembling each other (**Figure 6a** and **d**). Uniquely, the molar addition of N_3^- to **AS3** resulted in a slightly different spectra, comparing to the previous three (F^-, CN^-, OH^-), with the disappearance of the ICT band at 427 nm, simultaneously with the formation of new bands at 330 nm and 526 nm (**Figure 6b**). This was accompanied by isosbestic points observed at 308 nm, 373 nm and 466 nm respectively, signifying the formation of the pedant complex, and the co-existence of the probe and the complex at equilibrium. The interaction between N_3^- and **AS3** is still suspectedly through hydrogen bonding, even though it slightly differs from the rest, the binding position might be different.

Notingly, among the rest of the anions, $H_2PO_4^-$ was still able to induce changes when added to **A3**, same way like F^-, CN^- & OH^-, with all spectral characteristics (**Figure 7a**). Several isosbestic points were observed indicating that two species were co-existing in equilibrium. In addition, presence of AcO^- when molar added to **AS3**, spectral changes were observed, displaying similar behavior to **AS3-F**, **AS3-CN** and **AS3-OH** as indicated (**Figure 7f**). This illustrates that the interaction between **AS3** and AcO^- was of hydrogen bonding nature, through the hydroxyl groups of the

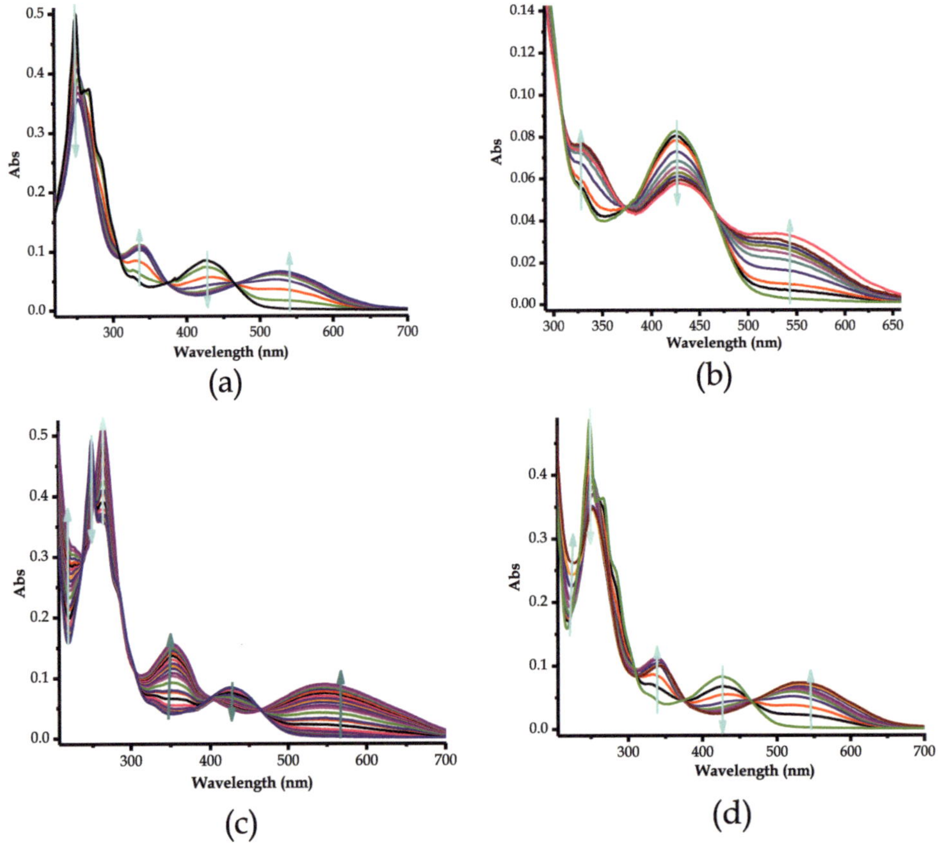

Figure 6.
The absorption titration spectra of **AS3** *(1 x 10*$^{-5}$ *M) in* CH_3CN*, with 5 equiv. of (a)* CN^-*, (b)* N_3^-*, (c)* F^- *and (d)* OH^- *at room temperature.*

probe. Other anions used could not induce any significant changes when added to **A3** (**Figure 7b** and **7c**) and **AS3** (**Figure 7d**) respectively. Thus, the main variation of the two probes (**A3** & **AS3**) towards anions was observed with the discrimination of N_3^-, which **AS3** was able to detect in addition to the other anions. Moreover, unlike **A3**, **AS3** was unable to detect the presence of $H_2PO_4^-$ as indicated (**Figure 7a**), probably due to the presence of the sulfonic acid group in **AS3**.

3.1.4 Absorption studies of A3 *and* AS3 *on interaction with cations*

Complementary to colorimetric experiments, UV–Vis spectroscopic experiments were conducted to investigate how absorption properties of **A3** and **AS3** were influenced by the presence of cations in the given solvent system. Subsequently, the molar addition of Cu^{2+} to **A3** resulted in spectra shifts of significant distinction. The molar titration of **A3** with Cu^{2+} resulted in the gradual disappearance of the ICT band at 423 nm concomitant with the appearance of a new band at 383 nm, accompanied by an isosbestic point at 383 nm (**Figure 8a**). The isosbestic point clearly shows the formation of a complex (**A3-Cu**) from the probe (**A3**), the mechanism attributable to electronic energy transfer caused by the coordination between the guest (Cu^{2+}) and the host (**A3**) species. The chelation-induced spectral changes were due to

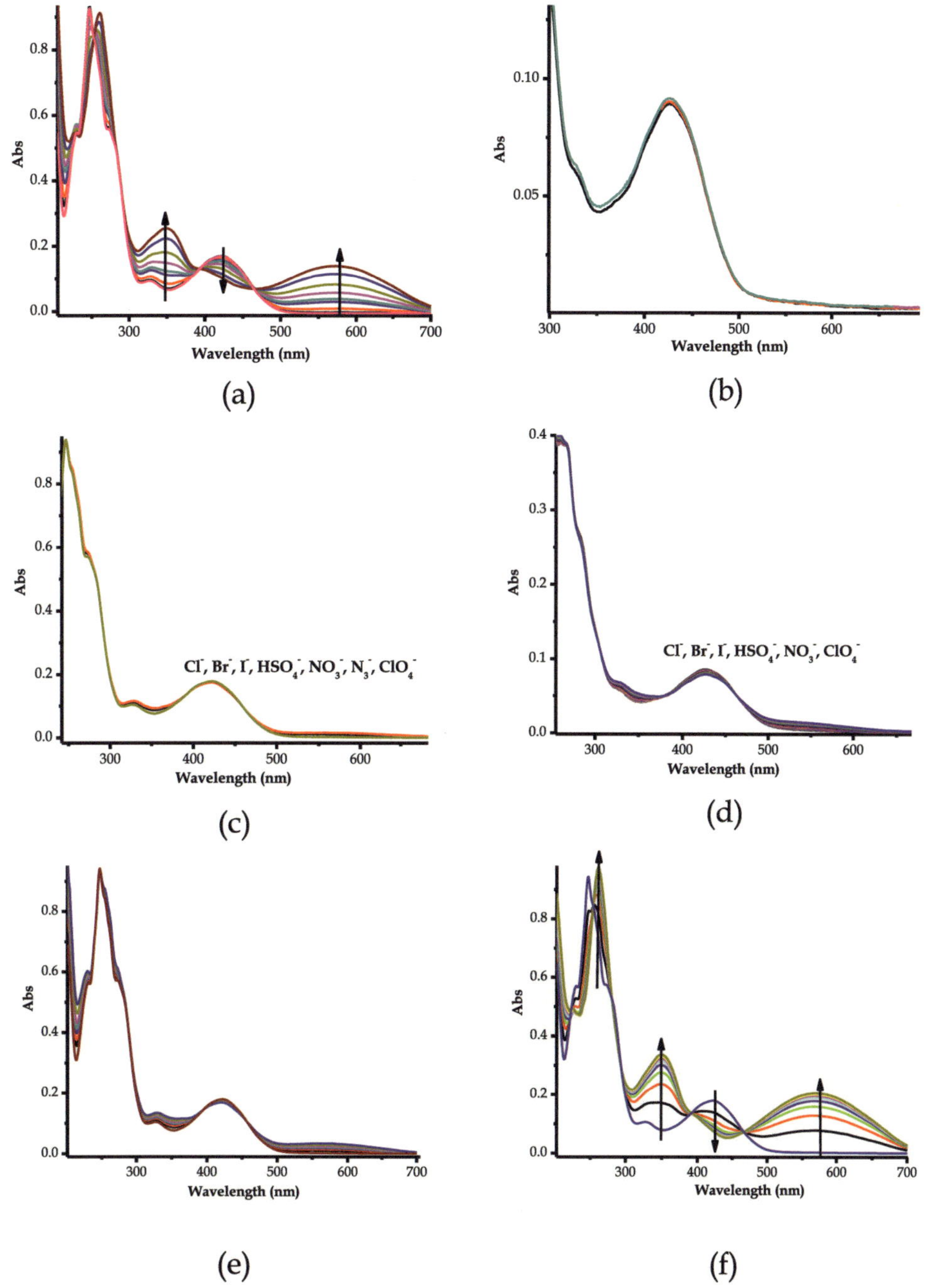

Figure 7.
The absorption titration spectra of **A3** *(1 x 10^{-5} M) in CH_3CN, with 5 equiv. of (a) $H_2PO_4^-$, (b) N_3^-, (c) other anions, and* **AS3** *with 5 equiv. of (d) other anions, (e) $H_2PO_4^-$, (f) AcO^-, at room temperature.*

the coordination of the p-orbitals of the oxygen atoms (hydroxyl and/or carbonyl group) of **A3** to the empty d-orbitals of Cu^{2+} to form a copper complex (**A3-Cu**). The chelation-induced change in spectra were in agreement with the light green color displayed upon introducing Cu^{2+} to **A3**, which was distinctively different from the interactions with other cation (**Figure 8**). However, the characteristics of the spectra

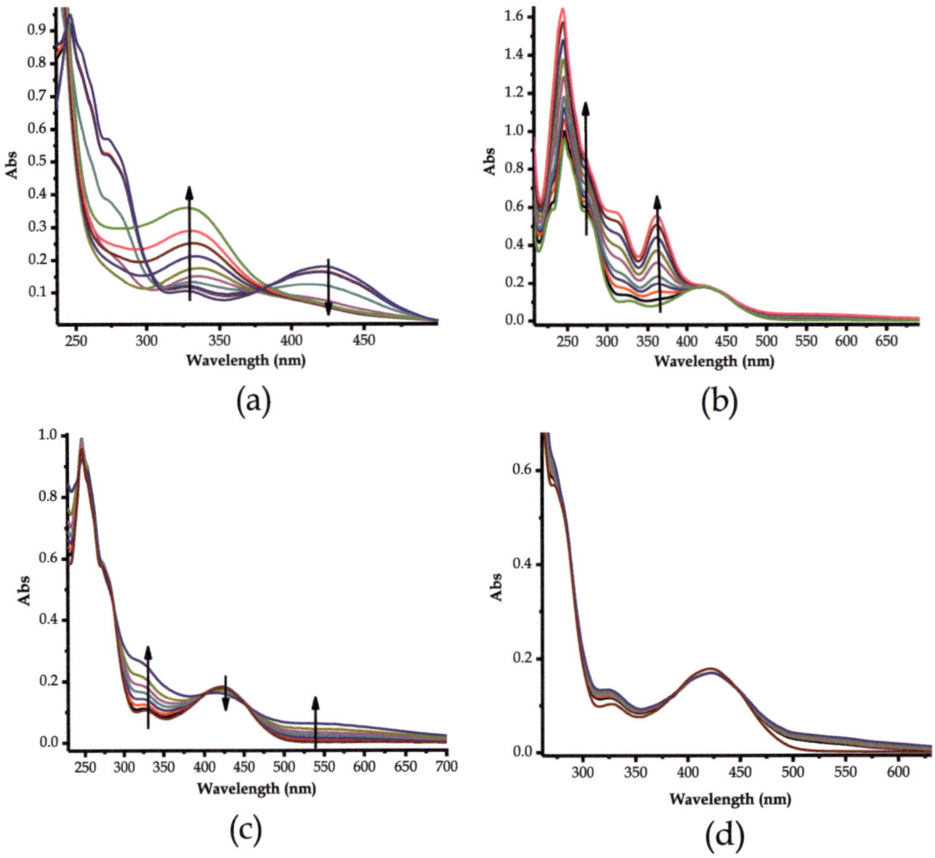

Figure 8.
The absorption titration spectra of **A3** *(1 x 10^{-5} M) in CH_3CN, with 5 equiv. of (a) Cu^{2+}, (b) Fe^{2+}, (c) Fe^{3+}, and (d) Zn^{2+} at room temperature.*

as a result of titration, are suggestive that a two-stage interaction was possible, where Cu^{2+} was interacting with **A3** in two different sites (in stages), given the geometrical identity of the molecule.

Furthermore, the only other cations that could induce significant changes when introduced to **A3** were Fe^{2+} and Fe^{3+}. For instance, the molar titration of Fe^{2+} with **A3** resulted in the emergence of two peaks in the UV-region at 309 nm and 360 nm, as well as the hyperchromic shift of the $\pi \rightarrow \pi^*$ transition band at 243 nm (**Figure 8b**). The formation of a new peak, complemented by the color change, was ascribed to the coordination of Fe^{2+} to **A3** forming a complex (**A3-Fe**). It was also noticed that no change was observed in the ICT band at 423 nm. Interestingly, the addition of Fe^{3+} was distinctively different, displaying a slight hypochromic shift of the ICT band at 423 nm, and the appearance of two new broader peaks in the regions of 300 nm to 360 nm, as well as 500 nm to 600 nm (**Figure 8c**), with two weakly identifiable isosbestic points, at 397 nm and 452 nm, respectively. The spectral behaviors are completely different from those of Fe^{2+}, which signify the possible paramagnetic (Fe^{3+}, d^5) and diamagnetic (Fe^{2+}, d^6) property influence for the two cations. The difference could also stem from the fact, the two have different atomic sizes (varying atomic radii), which could play a significant role into geometrical complementarity between the guest (Fe) and host (**A3**), in terms of shape and size, let alone the electrostatic force. The addition of other competitive cations, including Zn^{2+} could only induce slight changes (**Figure 8d**).

Contrastingly, the introduction of cations to **AS3** displayed slightly varying outputs as compared to **A3**. In addition to Cu^{2+}, Fe^{2+} and Fe^{3+} discriminated by **A3**, the probe (**AS3**) was able to detect the presence of other cations such as Ni^{2+} and Zn^{2+} in the same solvent medium. However, similarly to **A3**, the probe could discriminate the presence of Cu^{2+}, Fe^{2+} and Fe^{3+} in the same manner, with the disappearance of the peak at 423 nm, with the appearance of a new peak with maximum absorption at 299 nm upon the molar addition of Cu^{2+} (**Figure 9a**), with an isosbestic point at 388 nm. The characteristics of the spectra were still suspect that there exists a two-stage interaction between the guest (Cu^{2+}) and host (**AS3**), which translates into Cu^{2+} interacting with probe through **site 1**, before interacting again at **site 2** moment later, chronologically. Moreover, the molar titration of Fe^{2+}, resulted in more distinctive and intense peaks, such as the hyperchromic shift experienced by the $\pi \rightarrow \pi^*$ band at 240 nm, followed by two new peaks at 310 nm and 361 nm (**Figure 9b**). These peaks were followed by a new broader chelation-induced metal-to-ligand charge transfer (MLTC) band in the visible region (450 nm to 650 nm), with two weakly recognizable isosbestic points, at 422 nm and 440 nm, respectively.

Moreover, it was noticeable that the interaction behaviors of Fe^{3+}, Ni^{2+} and Zn^{2+} are of the same nature, based on the characteristics of their respective absorption spectra. The molar introduction of these cations to **AS3** displayed similarities in spectral characteristics, for example, all three had experience a new ICT band in the UV-region, a hypochromic shift at 425 nm and chelation-induced MLTC band in the visible region (**Figure 9c–e**). In both cases, three isosbestic points were observed, indicating the formation of complex pedants, upon interacting with the host species. In addition, other cations used did not induce any significant spectral shift, such Hg^{2+} as displayed (**Figure 9f**). It is worth noting, molecular identities of alizarin nature (**A3** & **AS3**) are highly likely to undergo ESIPT mechanism, which could describe varying characteristics of the formed complexes with different cations.

3.1.5 Selectivity studies of **A3** *and* **AS3** *on towards cations and anions*

The selectivity of **A3** towards cations observed was relatively indistinguishable among several cations (Zn^{2+}, Fe^{3+}, Ni^{2+}, Hg^{2+}), except for Cu^{2+} and Fe^{2+}, which displayed well-resolved spectra (**Figure 10a**). The addition of 5 equiv. each of cations demonstrated that **A3** was mostly selective only to Cu^{2+} and Fe^{2+}, however, in likely different modes of interaction, thus resulting in varying spectra. The color and spectral changes observed are resulting from a coordination induced charge transfer upon the interaction between the host (**A3**) and the guest (cations). Moreover, similar patterns were observed for **AS3** upon interacting with cations (**Figure 10b**). The combined addition of 5 equiv. each of cation to **AS3** saw Cu^{2+} and Fe^{2+} behaving differently from others, more similar to what was observed with **A3**, however, in addition Zn^{2+} and Ni^{2+} showed significant response in terms of spectral variation (**Figure 10b**). The sensitivity and selectivity variation of **AS3** over **A3** towards cations, displayed the effect of a sulfonic acid group ($-SO_3H$) has on the structure in terms of electronic transitions or charge transfers. Predictably, the interaction of **A3** and **AS3** is via coordination through the oxygen donor atoms (of hydroxyl groups and the ketones), thus with the additional oxygen donor atoms of the sulfonic acids, more coordination were possible. Therefore, recognition of Ni^{2+} and Zn^{2+} was observed, in addition to Cu^{2+} and Fe^{2+} for **AS3**, as compared to **A3**.

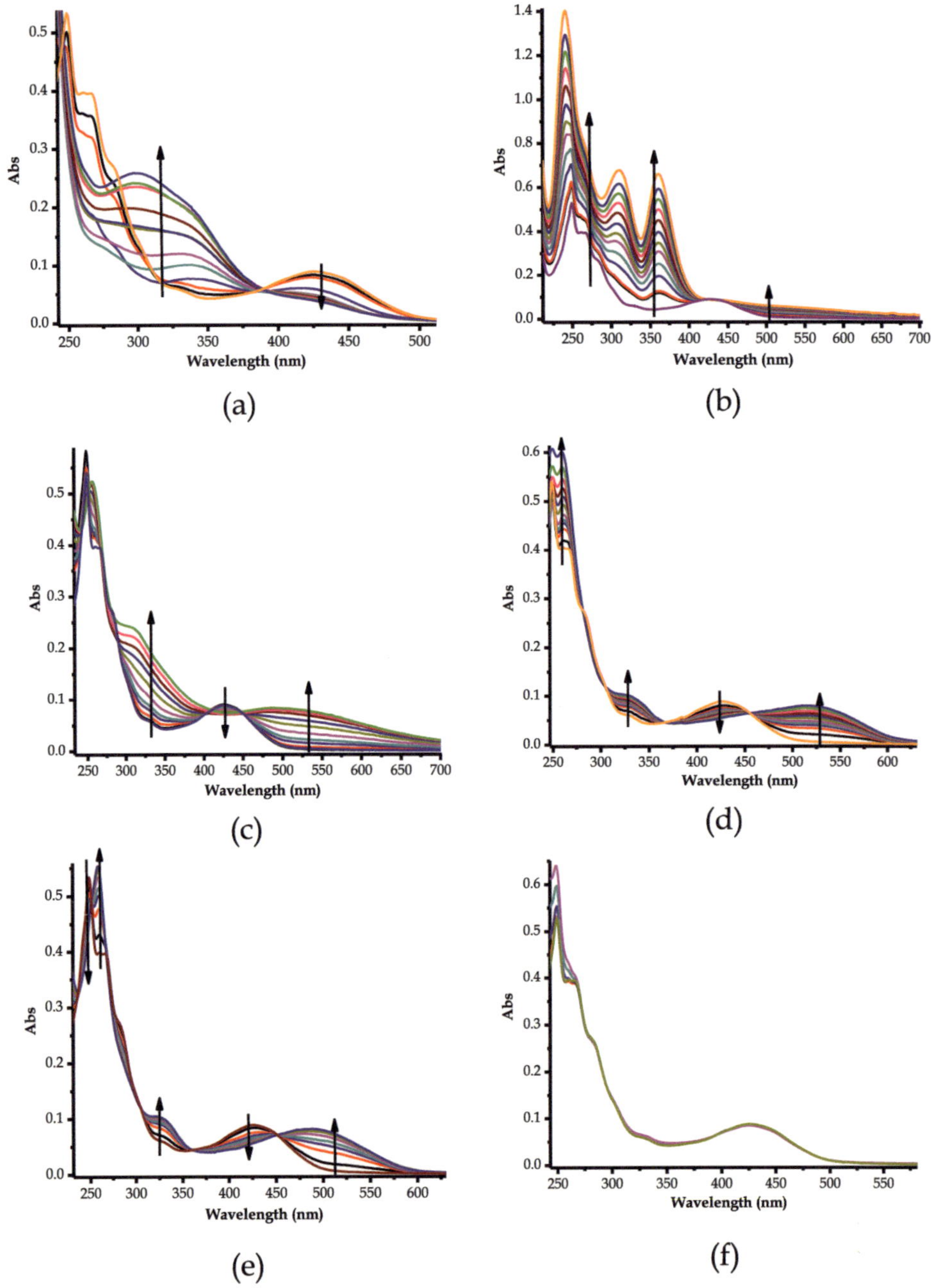

Figure 9.
The absorption titration spectra of **AS3** *(1 x 10*$^{-5}$ *M) in* CH_3CN*, with 5 equiv. of (a)* Cu^{2+}*, (b)* Fe^{2+}*, (c)* Fe^{3+}*, (d)* Ni^{2+} *(e)* Zn^{2+} *and (f)* Hg^{2+} *at room temperature.*

Furthermore, the two sensors were responsive commonly to four anions (F^-, CN^-, OH^-, AcO^-) as displayed above. However, upon the addition of 1 equiv. of each anion, the spectra intensities of AcO^- and OH^- were similarly high than all others for **A3** (**Figure 10c**), while for **AS3** it was OH^- and CN^- which were higher than others (**Figure 10d**). Thus, the information displayed is informative about the selectivity of

DOI: http://dx.doi.org/10.5772/intechopen.103829

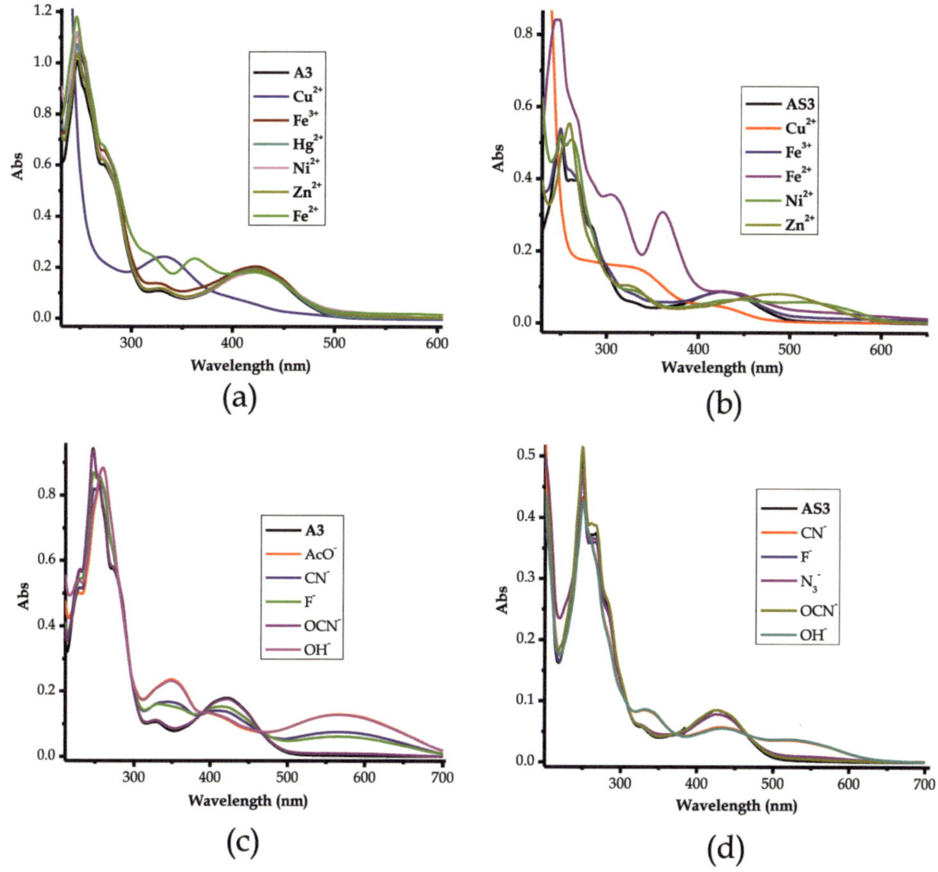

Figure 10.
The combined absorption titration spectra of (a) **A3** *(1 x 10^{-5} M) with 5 equiv. of each cation, (b)* **AS3** *(1 x 10^{-5} M) with 5 equiv. of each cations, (c)* **A3** *(1 x 10^{-5} M) with 1 equiv. of each anion and (d)* **AS3** *(1 x 10^{-5} M) with 1 equiv. of each anion, all in CH_3CN.*

the two sensors towards the anions. Evidently, OH^- and AcO^- exhibit high affinity towards **A3**, while OH^- and CN^- have highest binding affinity to **AS3**. The interaction of anions with the sensors is through hydrogen-bonding with the hydroxyl groups of **A3** and **AS3**, in most cases leading to deprotonation. The presence of the sulfonic acid group in **AS3** has harnessed the sensitivity the sensor to further discriminate CN^- in comparison to **A3**.

3.2 Fluorescence studies

In previous studies, an alizarin molecule (**A3**) has been found to exhibit fluorescence emission in the region of 600–620 nm, excited at 457 nm, using a range of selected solvents [4]. Thus, based on the existing data, fluorescence studies of **A3** and **AS3** were performed in CH_3CN, in order to compare and contrast the effect of cations and anions upon interaction, to complement information observed in absorption studies. Fluorescence analysis of **A3** has shown that the emission spectrum was characterized by a single moderate energy peak centered at 612 nm (λ_{ext} = 457 nm), while **AS3** was defined by a single emission peak at 600 nm (λ_{ext} = 440 nm). The structural variations of **A3** and **AS3** have resulted in slight emission spectral variation, hence different excitation wavelengths of the two structures. The blue shift exhibited by **AS3** as compared to **A3**, is attributed to the presence of a strong electron withdrawing group of sulfonyl group. In essence, there exists two possible charge transfer mechanisms in each structure, the locally excited state and the proton transfer state, which coexist. Mostly in such cases, the proton transfer state normally occurs faster, thereby offsetting the locally excited state one, resulting in a single peak [4], in this case exhibited by both **A3** and **AS3** moieties.

Upon the molar titration with cations, the emission spectrum of **A3** gradually experienced quenching process, varying from individual cations. For instance, the addition of Cu^{2+} (3 equiv.) resulted into a significant quenching process of more than 98% (**Figure 11a**), ascribed to the coordination interaction of **A3** with Cu^{2+} resulting in complex pedant (**A3-Cu**). Similar quenching trends were observed upon titration with Ni^{2+}, where the emission spectrum gradually decreased with increasing molar addition (**Figure 11b**) of the cation. Furthermore, the molar titration of **A3** with 3 equiv. of Cu^{2+}, resulted in 100% quenching effect (**Figure 11a**). Interestingly, the addition of Zn^{2+} displayed completely opposite behaviors of fluorescence enhancement (**Figure 11c**). Contrary to all other cations used, which have all resulted in emission quenching, the addition of Zn^{2+} yielded fluorescence enhancement, signifying

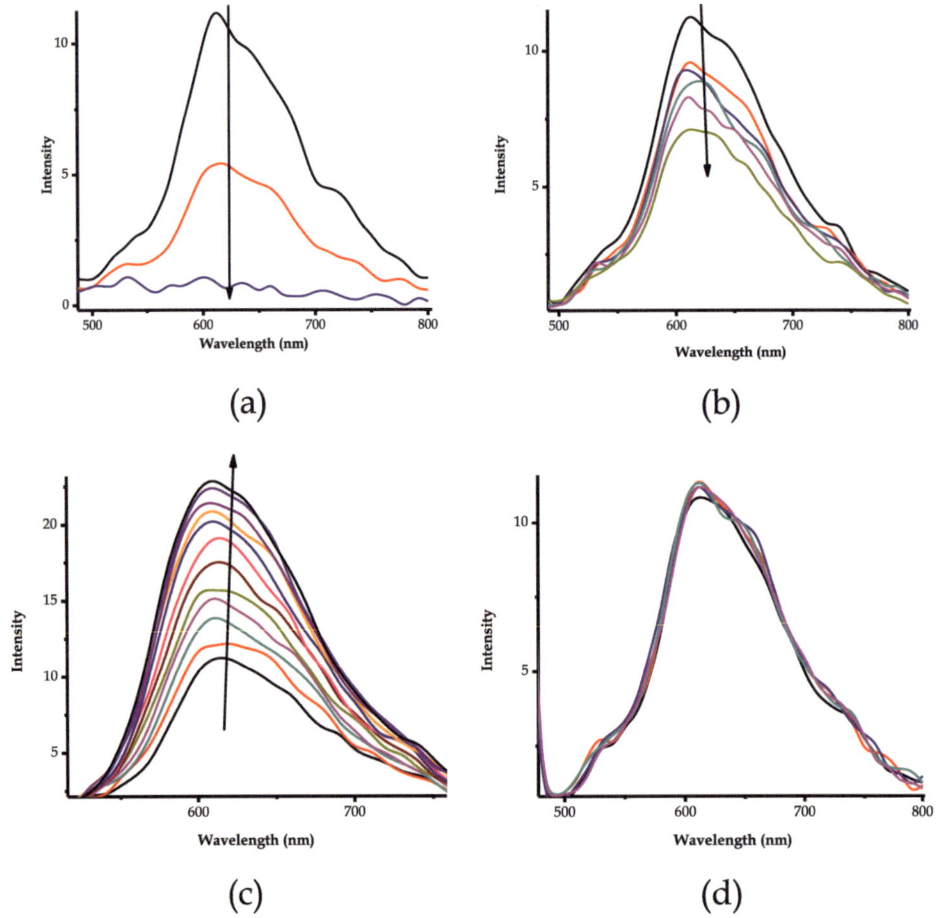

Figure 11.
The fluorescence titration spectra of **A3** *(1 x 10*$^{-5}$ *M) in* CH_3CN*, with (a)* Cu^{2+} *(3 equiv.), (b)* Ni^{2+} *(20 equiv.), (c)* Zn^{2+} *(30 equiv.), and (d)* Fe^{3+} *(20 equiv.), at* λ_{ext} *= 457 nm.*

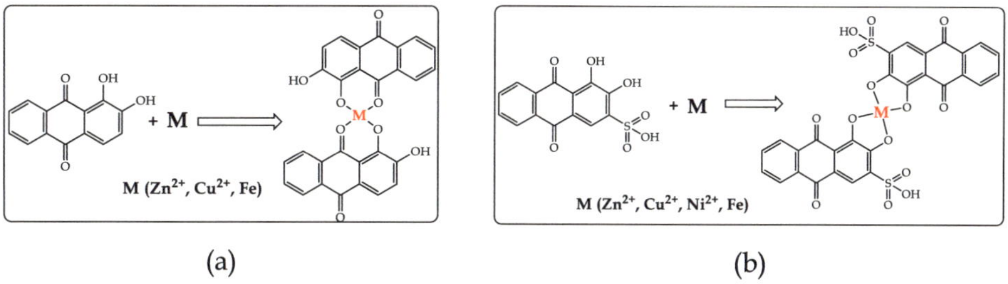

Figure 12.
Proposed binding mechanisms for (a) **A3** *with cations, and (b)* **AS3** *with cations*

that the nature of interaction was perhaps different from those of other cations. The quenching effect was suspectedly due to the combination of chelation-enhanced fluorescence (CHEF) and internal charge transfer (ICT) mechanisms. The addition of other cations, including Fe^{3+} did not induce any significant change to the spectrum (**Figure 11d**). Generally, the interaction modes of cations towards **A3** and **AS3** are via coordination through hydroxyl oxygen donor atoms within the structures [42–51]. It has also been established that the binding modes of particularly Zn^{2+} and Cu^{2+} among other transition metals towards **A3** and/or **AS3** entities is via hydroxyl groups [42, 43, 46, 47, 50], inducing ESIPT process, in a 2:1 interaction ratio (**Figure 12a** and **b**).

Furthermore, the interaction of cations with **AS3** showed closely similar results as in **A3**, however, a distinctive variance was observed in the association with Zn^{2+} and Fe^{3+}. The addition Cu^{2+} to **AS3** resulted in a 100% quenching process, showing that a strong association of **AS3-Cu** (**Figure 13a**), more like in **A3-Cu** pedant. A similar trend was observed upon the addition of Ni^{2+} to **AS3**, where 63% quenching was observed (**Figure 13c**). Interestingly, unlike in **A3**, the addition of Zn^{2+} to **AS3** resulted in a significant fluorescence quenching (**Figure 13b**), with a quenching of 83% attained after adding 10 equiv. The different behavior of **AS3-Zn** is ascribed to the presence of a strong electron withdrawing sulfonyl group, which has a suppressing or disruptive effect on charge transfer mechanisms, the coordination-based charge transfer and the ICT. Moreover, another new trend observed for **AS3**, was the interaction with Fe^{3+}, unlike in **A3**, the molar addition of Fe^{3+} to **AS3** resulted in a significant quenching of 67% (**Figure 13d**) due to chelation effect, even though about 20 equiv. The interaction modes of cations with **AS3** were similar to **A3**, through coordination with the hydroxyl groups of the **AS3**, mostly in a 2:1 interaction ratio (**Figure 12a** and **b**).

The interaction of biological anions with **A3** and **AS3** were studied fluorometrically, in CH_3CN. The molar addition of anions (F^-, CN^-, OH^-, AcO^- and N_3^-) resulted in fluorescence quenching (**Figure 14**). However, the incremental addition of F^-, CN^-, OH^- and AcO^- to **A3**, suggested a two-step interaction behavior, where initial molar addition resulted in obvious fluorescence quenching at 612 nm until a certain quantity was added, then interestingly, the continual addition suddenly gave rise to fluorescence enhancement (**Figure 14a-d**) with a new peak at 656 nm. The first step of fluorescence quenching was ascribed to hydrogen bonding interaction between anions and **A3**, while the subsequent enhancement observed shortly, with incremental excessive addition of the anions, was attributed to the deprotonation effect. The divorce of the hydrogen ion off the structure of **A3** via ESIPT mechanism, resulted in the emissive properties of **A3** restored, hence a new fluorescence enhancement spectra at a different wavelength (656 nm), with increasing addition of anions. The new emission peak at 656 nm

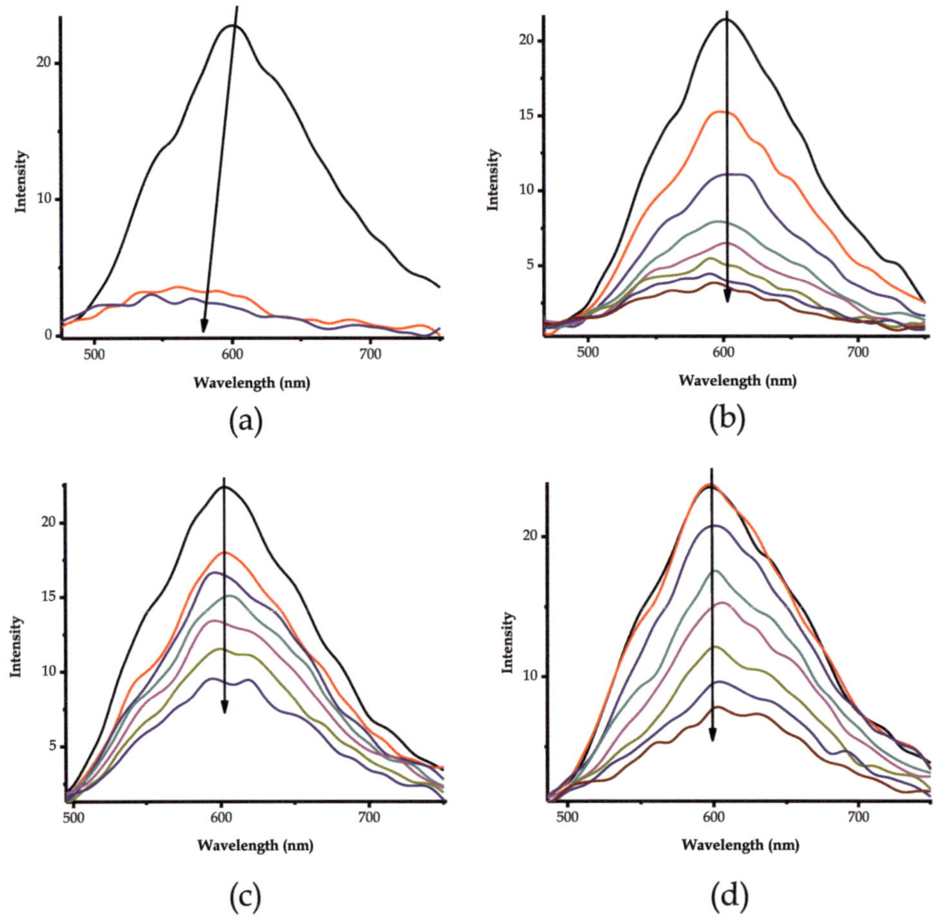

Figure 13.
The fluorescence titration spectra of **AS3** *(1 x 10^{-5} M) in CH_3CN, with (a) Cu^{2+} (3 equiv.), (b) Zn^{2+} (10 equiv.), (c) Ni^{2+} (20 equiv.), and (d) Fe^{3+} (20 equiv.), at λ_{ext} = 440 nm.*

signifies a change in the structure upon the removal of a hydrogen ion through deprotonation, thus red shifting the spectra from 612 nm (**A3**) to 656 nm (deprotonated **A3**). The other anion which induced fluorescence quenching upon interacting with **A3**, was N_3^- (**Figure 14e**), however, no deprotonation effect was suspect, even when large quantities were added, ascribe to perhaps weaker interactions with **A3**.

Moreover, the effect of an electron withdrawing sulfonyl group was apparent from the activities of emission spectra of **AS3** upon interaction with anions (F^-, CN^-, OH^-, AcO^- and N_3^-). The interactions of all anions used induced significant fluorescence quenching (**Figure 15a-e**) due to strong association with **AS3**, through hydrogen bonding mechanism, however, no deprotonation process seemed to have taken place. Unlike in **A3** where deprotonation was observed, **AS3** did not display any signal of change in fluorescence behavior (red or blue shift), apart from quenching. The difference between the two probes (**A3** and **AS3**) was the sulfonyl group, which is a strong electron-withdrawing species, thereby polarizing the molecule, thus inhibiting fluorescence enhancement due to the deprotonation activity. It is obvious, the addition of 10 equiv. of the anions resulted in the decrease of fluorescence spectrum

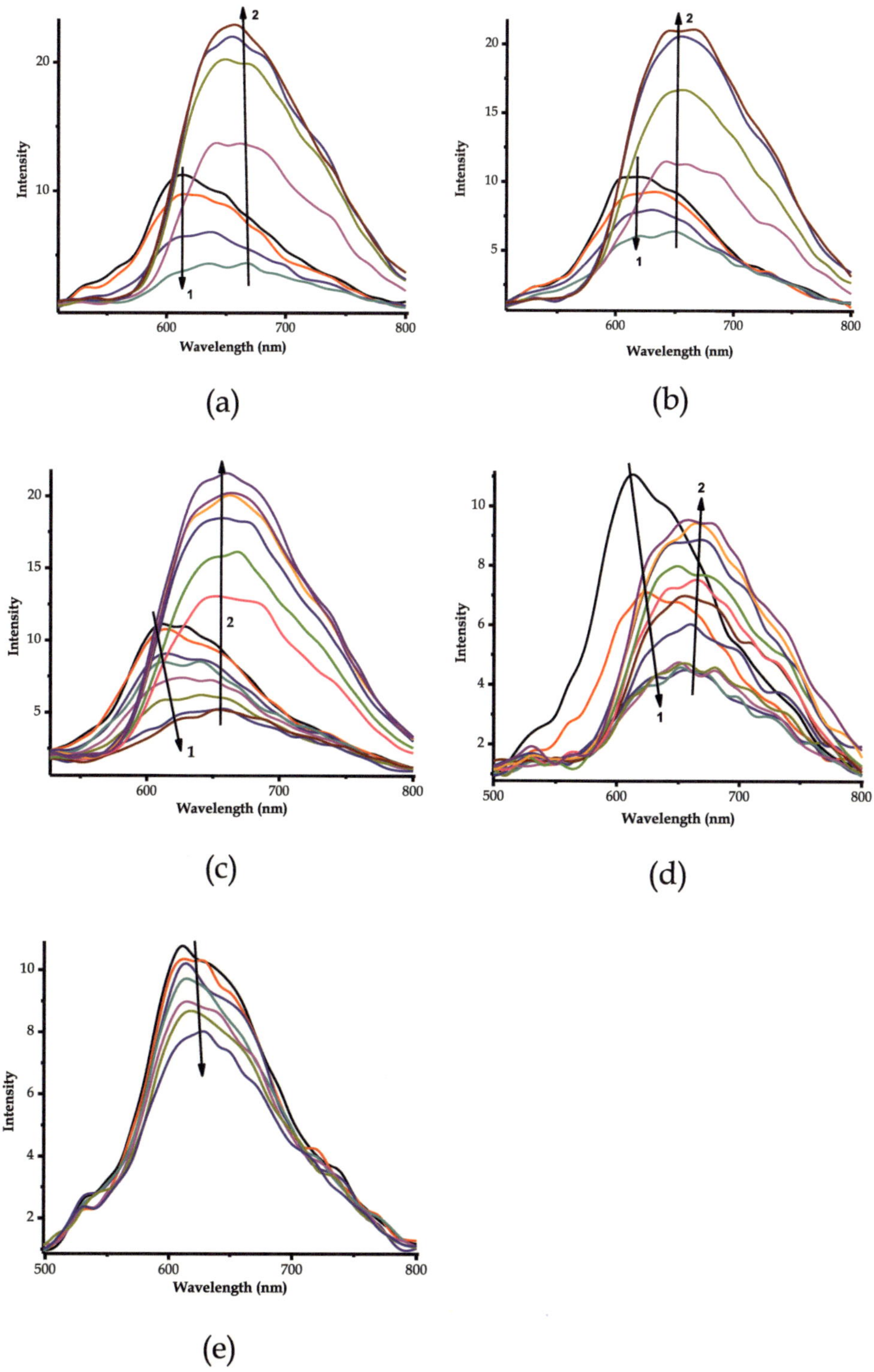

Figure 14.
The fluorescence titration spectra of **A3** *(1 x 10^{-5} M) in CH_3CN, with (a) F^- (20 equiv.), (b) OH^- (10 equiv.), (c) CN^- (20 equiv.), (d) AcO^- (40 equiv.) and (e) N_3^- (30 equiv.), at λ_{ext} = 457 nm.*

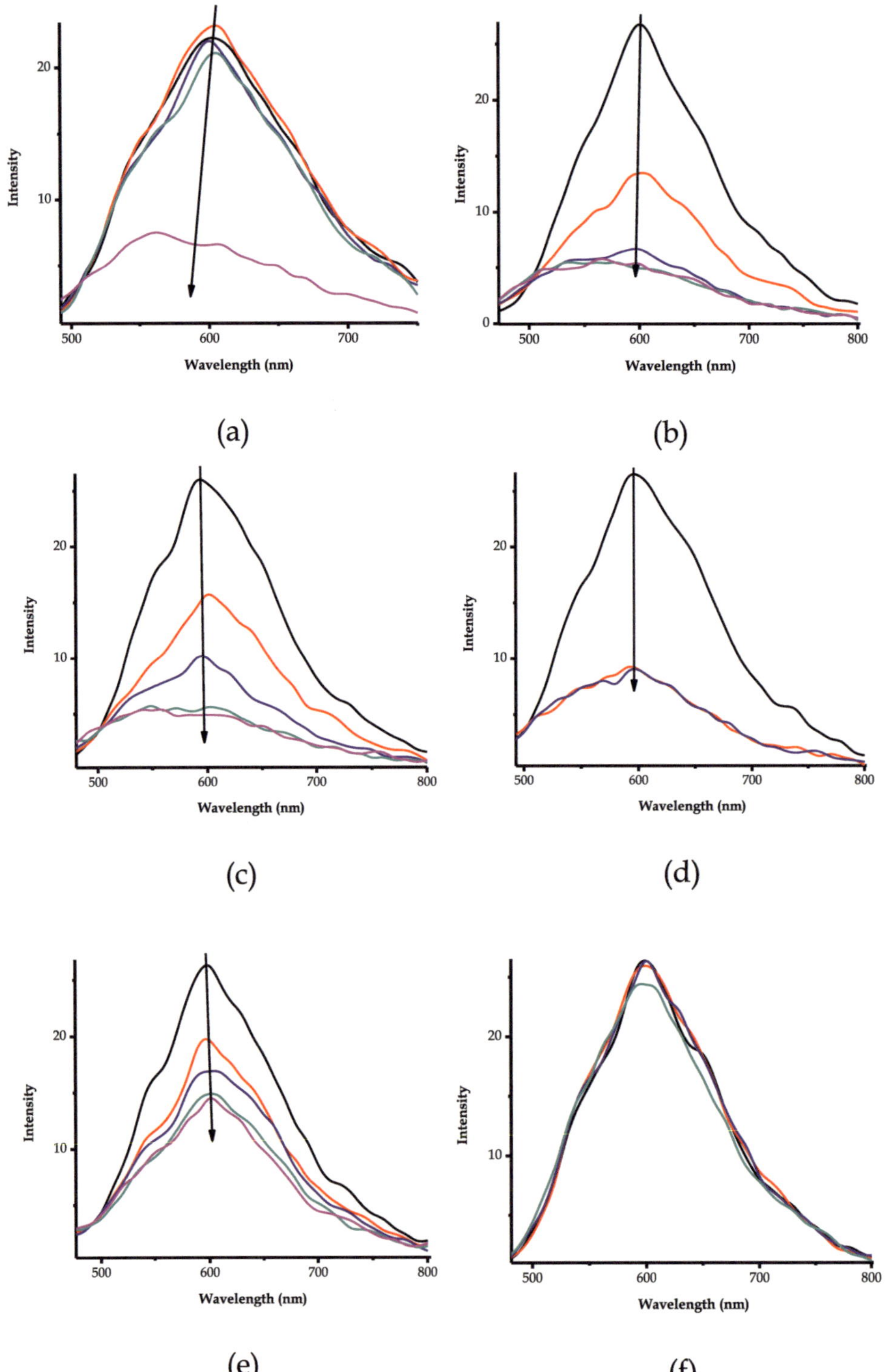

Figure 15.
The fluorescence titration spectra of **AS3** *(1 x 10^{-5} M) in CH_3CN, with (a) F^- (10 equiv.), (b) CN^- (10 equiv.), (c) OH^- (10 equiv.), (d) AcO^- (5 equiv.), (e) N_3^- (30 equiv.) and (f) $H_2PO_4^-$ (50 equiv.), at λ_{ext} = 440 nm.*

at 612 nm, with the quenching strength depending on the nature of anions, F^- (75% quenching), CN^- (79% quenching), OH^- (79% quenching) and N_3^- (79% quenching). There was no significant fluorescence change signals upon the addition of other anions, including $H_2PO_4^-$ (**Figure 15f**).

4. Conclusion

Conclusively, the comparative studies on the chemosensing property studies of the two Alizarin probes (**A3** and **AS3**) were successfully conducted. The two entities displayed closely similar behaviors towards the discrimination of cations and anions in CH_3CN. However, variations in behaviors was observed in cations towards **A3** and **AS3**, where each probe was able detect certain cations, by means of UV–Vis absorption titrations, as well as fluorometric responses. The UV–Vis titration of **A3** with cations resulted in selective detection of Cu^{2+}, Fe^{3+}, Fe^{2+} and Zn^{2+}, while **AS3** was responsive towards Cu^{2+}, Fe^{3+}, Fe^{2+}, Ni^{2+} and Zn^{2+} respectively. The behaviors of anions towards **A3** and **AS3** in UV–Vis titrations (so as colorimetrically) and fluorometrically were consistently similar. On the hand, fluorometric titrations resulted in distinctive behaviors, where mostly fluorescence quenching effect was observed upon adding cations to **A3** or **AS3** in CH_3CN. However, distinctive features were experienced upon adding Zn^{2+} to **A3** and **AS3**, where chelation fluorescence enhancement (CHEF) and chelation enhanced fluorescence quenching (CHEQ) were observed, respectively. Another distinctive feature between the two probes observed was the deprotonation effects exhibited by **A3** upon interacting with anions (F^-, CN^-, OH^- and AcO^-), where at first a quenching effect was observed, then further addition resulted in the enhancement effect (due to deprotonation effect). However, this was not the case for the with **AS3**, where all anions were restricted to fluorescence quenching ONLY, even after large quantities were added. The differences in fluorometric properties of **A3** and **AS3** towards Zn^{2+} and anions, were influenced by the presence of the sulfonyl group in **AS3**, which was the main determining factor, since it is a strong electron withdrawing group. Thus, it can be said that the two probes have displayed interesting features in applications towards the detection of cations and anions, which can be used in the development of analytical probes for real time applications.

Acknowledgements

This work was supported by the Department of Physics, Chemistry & Material Science, University of Namibia, Namibia. The work was also partly supported by the Royal Society-DFiD Africa Capacity Building Initiative, New Materials for a Sustainable Energy Future.

Conflicts of interest

There are no conflicts of interest to declare.

Author details

Veikko Uahengo
Department of Physics, Chemistry and Material Science, University of Namibia, Windhoek, Namibia

*Address all correspondence to: vuahengo@gmail.com

References

[1] Vankar PS, Shanker R, Mahanta D, Tiwari SC. Ecofriendly sonicator dyeing of cotton with Rubia cordifolia Linn. Using biomordant. Dyes and Pigments. 2008;**76**(1):207-212

[2] Bien H-S, Stawitz J, Wunderlich K. Anthraquinone dyes and intermediates. Ullmann's Encyclopedia of Industrial Chemistry. 2000;**3**:515-578. DOI: 10.1002/14356007.a02_355

[3] Zhang MX, Zhao GJ. Dimerization assembly mechanism involving proton coupled electron transfer for hydrogen evolution from water by molybdenum-oxo catalyst. Journal of Alloys and Compounds. 2016;**664**:439-443. DOI: 10.1016/j.jallcom.2016.01.014

[4] Sasirekha V, Umadevi M, Ramakrishnan V. Solvatochromic study of 1,2-dihydroxyanthraquinone in neat and binary solvent mixtures. Spectrochimica Acta Part A: Molecular and Biomolecular Spectroscopy. 2008;**69**(1):148-155

[5] Takahashi E, Fujita KI, Kamataki T, Arimoto-Kobayashi S, Okamoto K, Negishi T. Inhibition of human cytochrome P450 1B1, 1A1 and 1A2 by antigenotoxic compounds, purpurin and alizarin. Mutation Research, Fundamental and Molecular Mechanisms of Mutagenesis. 2002;**508**(1-2):147-156

[6] Kaniyankandy S, Verma S, Mondal JA, Palit DK, Ghosh HN. Evidence of multiple electron injection and slow back electron transfer in alizarin-sensitized ultrasmall TiO2 particles. Journal of Physical Chemistry C. 2009;**113**(9):3593-3599

[7] Duncan WR, Prezhdo OV. Theoretical studies of photoinduced electron transfer in dye-sensitized TiO2. Annual Review of Physical Chemistry. 2007;**58**:143-184

[8] Huber R, Moser JE, Grätzel M, Wachtveitl J. Real-time observation of photoinduced adiabatic electron transfer in strongly coupled dye/semiconductor colloidal systems with a 6 fs time constant. The Journal of Physical Chemistry. B. 2002;**106**(25):6494-6499

[9] Le Person A, Cornard JP, Say-Liang-Fat S. Studies of the tautomeric forms of alizarin in the ground state by electronic spectroscopy combined with quantum chemical calculations. Chemical Physics Letters. 2011;**517**(1-3):41-45

[10] Fotia C, Avnet S, Granchi D, Baldini N. The natural compound alizarin as an osteotropic drug for the treatment of bone tumors. Journal of Orthopaedic Research. 2012;**30**(9):1486-1492

[11] Habeeb MM, Alghanmi RM. Spectrophotometric study of intermolecular hydrogen bonds and proton transfer complexes between 1,2-dihydroxyanthraquinone and some aliphatic amines in methanol and acetonitrile. Journal of Chemical & Engineering Data. 2010;**55**(2):930-936

[12] Trotsek D. No title No title. Journal of Chemical Information and Modeling. 2017;**110**(9):1689-1699

[13] Mech J, Grela MA, Szaciłowski K. Ground and excited state properties of alizarin and its isomers dedicated to professor Krzysztof Fitzner of the occasion of his 70th birthday. Dyes and Pigments. 2014;**103**:202-213

[14] Kim YC, Lee SH, Kim MS, Cha YB, Ahn KH. Electroluminescence characteristics of a new green-emitting phenothiazine derivative with biphenyl benzimidazole substituent.

Molecular Crystals and Liquid Crystals. 2010;**520**:36-43

[15] Zhao J, Yao H, Liu J, Hoffmann MR. New excited-state proton transfer mechanisms for 1,8-dihydroxydibenzo[a,h]phenazine. The Journal of Physical Chemistry. A. 2015;**119**(4):681-688

[16] Das R, Klymchenko AS, Duportail G, Mély Y. Excited state proton transfer and solvent relaxation of a 3-hydroxyflavone probe in lipid bilayers. The Journal of Physical Chemistry. B. 2008;**112**(38):11929-11935

[17] Lim S-J, Seo J, Park SY. Photochromic switching of excited-state intramolecular proton-transfer (ESIPT) fluorescence: A unique route to high-contrast memory switching and nondestructive readout. Journal of the American Chemical Society. 2006;**128**:14542-14547

[18] Hong KI, Park SH, Lee SM, Shin I, Jang WD. A pH-sensitive excited state intramolecular proton transfer fluorescent probe for imaging mitochondria and helicobacter pylori. Sensors and Actuators B: Chemical. 2019;**286**:148-153. DOI: 10.1016/j.snb.2019.01.101

[19] Dahal D, McDonald L, Bi X, Abeywickrama C, Gombedza F, Konopka M, et al. An NIR-emitting lysosome-targeting probe with large stokes shift: Via coupling cyanine and excited-state intramolecular proton transfer. Chemical Communications. 2017;**53**(26):3697-3700. DOI: 10.1039/C7CC00700K

[20] Tasch S, List EJW, Ekström O, Graupner W, Leising G, Schlichting P, et al. Efficient white light-emitting diodes realized with new processable blends of conjugated polymers. Applied Physics Letters. 1997;**71**(20):2883-2885

[21] Santos FS, Ramasamy E, Ramamurthy V, Rodembusch FS. Excited state chemistry of flavone derivatives in a confined medium: ESIPT emission in aqueous media. Photochemical & Photobiological Sciences. 2014;**13**(7):992-996

[22] Jadhav AG, Sekar N. Substituent modulation from ESIPT to ICT emission in Benzoimidazolphenyl-methanones derivatives: Synthesis, Photophysical and DFT study. Journal of Solution Chemistry. 2017;**46**(4):777-797

[23] Kuzu B, Tan M, Ekmekci Z, Menges N. A novel structure for ESIPT emission: Experimental and theoretical investigations. Journal of Photochemistry and Photobiology A: Chemistry. 2019;**381**:111874. DOI: 10.1016/j.jphotochem.2019.111874

[24] Shigemitsu Y, Mutai T, Houjou H, Araki K. Excited-state intramolecular proton transfer (ESIPT) emission of hydroxyphenylimidazopyridine: Computational study on enhanced and polymorph-dependent luminescence in the solid state. The Journal of Physical Chemistry. A. 2012;**116**(49):12041-12048

[25] Wu L, Liu L, Han HH, Tian X, Odyniec ML, Feng L, et al. ESIPT-based fluorescence probe for the ratiometric detection of superoxide. New Journal of Chemistry. 2019;**43**(7):2875-2877

[26] Massue J, Pariat T, Vérité PM, Jacquemin D, Durko M, Chtouki T, et al. Natural born laser dyes: Excited-state intramolecular proton transfer (ESIPT) emitters and their use in random lasing studies. Nanomaterials. 2019;**9**(8):1-13. DOI: 10.3390/nano9081093

[27] Saravanan A, Shyamsivappan S, Kalagatur NK, Suresh T, Maroli N, Bhuvanesh N, et al. Application of real sample analysis and biosensing: Synthesis

of new naphthyl derived chemosensor for detection of Al3+ ions. Spectrochimica Acta Part A: Molecular and Biomolecular Spectroscopy. 2020;**241**:118684. DOI: 10.1016/j.saa.2020.118684

[28] Chen D, Chen P, Zong L, Sun Y, Liu G, Yu X, et al. Colorimetric and fluorescent probes for real-time naked eye sensing of copper ion in solution and on paper substrate. Royal Society Open Science. 2017;**4**(11):1-8. DOI: 10.6084/m9.figshare.c.3918079.v1

[29] Asaithambi G, Periasamy V. Hydrogen sulfide detection by ESIPT based fluorescent sensor: Potential in living cells imaging. Journal of Photochemistry and Photobiology A: Chemistry. 2019;**369**:97-105. DOI: 10.1016/j.jphotochem.2018.10.013

[30] Gupta VK, Mergu N, Kumawat LK, Singh AK. A reversible fluorescence "off-on-off" sensor for sequential detection of aluminum and acetate/fluoride ions. Talanta. 2015;**144**:80-89. DOI: 10.1016/j.talanta.2015.05.053

[31] Mathivanan M, Tharmalingam B, Lin CH, Pandiyan BV, Thiagarajan V, Murugesapandian B. ESIPT-active multi-color aggregation-induced emission features of triphenylamine-salicylaldehyde-based unsymmetrical azine family. CrystEngComm. 2019;**22**(2): 213-228

[32] Patel SM, Pal K, Kumar PN, Deepa M, Sharada DS. Design and synthesis of novel indole and Carbazole based organic dyes for dye sensitized solar cells: Theoretical studies by DFT/TDDFT. ChemistrySelect. 2018;**3**(6):1623-1628

[33] Huang D, Bing Y, Yi H, Hong W, Lai C, Guo Q, et al. An optical-fiber sensor based on time-gated fluorescence for detecting water content in organic solvents. Analytical Methods. 2015;**7**(11): 4621-4628. DOI: 10.1039/C5AY00110B

[34] Ishibashi Y, Murakami M, Araki K, Mutai T, Asahi T. Excited-state intramolecular proton-transfer process of crystalline 6-Cyano-2-(2′-hydroxyphenyl)imidazo[1,2 a]pyridine, as revealed by femtosecond pump-probe microspectroscopy. Journal of Physical Chemistry C. 2019;**123**(17): 11224-11232

[35] Halder D, Mallick A, Purkayastha P. Doping dopamine in carbon nanoparticles: A new multifunctional logic-based decision-making molecule. Langmuir. 2019;**35**(33):10885-10889

[36] Uahengo V, Xiong B, Cai P, Daniel LS, Rhyman L, Ramasami P. Chromogenic signaling of water traces by 1,8-naphthalohydrazone-anion complex in organic solvents. Analytical Chemistry Research. 2016;**8**:1-8. DOI: 10.1016/j.ancr.2016.03.001

[37] Tsumura S, Enoki T, Ooyama Y. A colorimetric and fluorescent sensor for water in acetonitrile based on intramolecular charge transfer: D-(π-A)2-type pyridine-boron trifluoride complex. Chemical Communications. 2018;**54**(72):10144-10147

[38] Miyaji H, Kim HK, Sim EK, Lee CK, Cho WS, Sessler JL, et al. Coumarin-strapped calix [4]pyrrole: A fluorogenic anion receptor modulated by cation and anion binding. Journal of the American Chemical Society. 2005;**127**(36):12510-12512

[39] Sathyaraj G, Muthamilselvan D, Kiruthika M, Weyhermüller T, Nair BU. Ferrocene conjugated imidazolephenols as multichannel ditopic chemosensor for biologically active cations and anions. Journal of Organometallic Chemistry. 2012;**716**:150-158

[40] Huang YH, Geng QX, Jin XY, Cong H, Qiu F, Xu L, et al. Tetramethylcucurbit[6]uril-triggered fluorescence emission and its application for recognition of rare earth cations. Sensors and Actuators B: Chemical. 2017;**243**:1102-1108. DOI: 10.1016/j.snb.2016.12.102

[41] Qiu S, Gao S, Zhu X, Lin Z, Qiu B, Chen G. Development of ultra-high sensitive and selective electrochemiluminescent sensor for copper(II) ions: A novel strategy for modification of gold electrode using click chemistry. Analyst. 2011;**136**(8):1580-1585 Available from: http://www.ncbi.nlm.nih.gov/pubmed/21331394

[42] Zhang L, Dong S, Zhu L. Fluorescent dyes of the esculetin and alizarin families respond to zinc ions ratiometrically. Chemical Communications. 2007;**19**:1891-1893

[43] Wu FY, Liao WS, Wu YM, Wan XF. Spectroscopic determination of cysteine with alizarin red S and copper. Spectroscopy Letters. 2008;**41**(8):393-398

[44] Ritter J, Borst H-U, Lindner T, Hauser M, Brosig S, Bredereck K, et al. Substituent effects on triplet yields in aminoanthraquinones: Radiationless deactivation via intermolecular and intramolecular hydrogen bonding. Journal of Photochemistry and Photobiology A: Chemistry. 1988;**41**:227-244

[45] Kunkely H, Vogler A. Fluorescence of alizarin complexone and its metal complexes. Inorganic Chemistry Communications. 2007;**10**(3):355-357

[46] Kaushik R, Kumar P, Ghosh A, Gupta N, Kaur D, Arora S, et al. Alizarin red S-zinc(II) fluorescent ensemble for selective detection of hydrogen sulphide and assay with an H2S donor. RSC Advances. 2015;**5**(97):79309-79316

[47] Montaseri H, Yousefinejad S. Design of an optical sensor for the determination of cysteine based on the spectrophotometric method in a triacetylcellulose film: PC-ANN application. Analytical Methods. 2014;**6**(21):8482-8487

[48] Zhu L, Bai YL, Zhao Y, Xing F, Li MX, Zhu S. Bis (2-pyridylmethyl) amine-functionalized alizarin: An efficient and simple colorimetric sensor for fluoride and a fluorescence turn-on sensor for Al 3+ in an organic solution. Dalton Transactions. 2019;**48**(15):5035-5047

[49] Yin C, Zhang J, Huo F. Combined spectral experiment and theoretical calculation to study the interaction of 1,4-dihydroxyanthraquinone for metal ions in solution. Spectrochimica Acta, Part A: Molecular and Biomolecular Spectroscopy 2013; 115: 772-777. Available from: doi:10.1016/j.saa.2013.06.095

[50] Sinha S, Gaur P, Mukherjee T, Mukhopadhyay S, Ghosh S. Exploring 1,4-dihydroxyanthraquinone as long-range emissive ratiometric fluorescent probe for signaling Zn2+/PO43-: Ensemble utilization for live cell imaging. Journal of Photochemistry and Photobiology B: Biology. 2015;**148**:181-187

[51] Doskocz M, Kubas K, Frackowiak A, Gancarz R. NMR and ab initio studies of Mg2+, Ca2+, Zn2+, Cu2+ alizarin complexes. Polyhedron. 2009;**28**(11):2201-2205

Chapter 6

Perspective Chapter: Slowing Down the "Internal Clocks" of Atoms – A Novel Way to Increase Time Resolution in Time-Resolved Experiments through Relativistic Time Dilation

Hazem Daoud

Abstract

Traditional time-resolved studies typically rely on a pump laser beam that triggers a reaction dynamic in an atom or molecule and is subsequently probed by a probe pulse of photons, electrons or neutrons. This traditional method is reliant on advancements in creating ever shorter probe and pump pulses. The shorter the pulses the higher is the time resolution. In this chapter we would like to present a novel idea that has the potential to achieve 2–3 orders of magnitude higher time resolutions than is possible with laser and electron compression technology. The proposed novel method is to slow down the 'internal clock' of the sample. This can be achieved by accelerating the sample to relativistic speeds, which can be realized in particle accelerators such as cyclotrons and synchrotrons.

Keywords: ultrafast science, femtosecond, attosecond, spectroscopy, electron diffraction, molecular dynamics, special relativity

1. Introduction

Until a few decades ago capturing molecular dynamics was in the realm of Gedanken- or thought experiments [1, 2]. Chemists know about the reactants and the final products of chemical reactions, but how the molecules and atoms rearrange themselves to produce the reaction products had always remained in the realm of imagination. This is due to the technical difficulties in making these measurements. In solids, chemical reactions occur at the speed of sound ($\sim$1000 m/s) and atomic bond lengths are on the order of 1 Å, which means that the time resolution required is on the order of femtoseconds [3, 4].

Recent advances in laser technology have made it possible to produce laser pulses that are femtoseconds and even attoseconds in duration [5]. This has enabled rapid developments in the field of ultrafast science. Typically, a short laser pulse initiates a photo-induced reaction dynamic in a molecular sample, which is then probed by a probe pulse. Probe pulses can be short X-ray pulses [6] in X-ray free electron lasers (XFELs) [7] or compressed electron pulses [8] in tabletop experiments [3, 9]. Additionaly, laser pulses can be used in table-top spectroscopy experiments, which temporally probe molecules and atoms but not with the same spatial resolution as X-rays [10, 11]. XFELs are multibillion dollar facilities that are very costly to operate and entail very complex engineering [12]. However, electron beams are generated in table-top experiments. The electrons are usually accelerated via a DC electric field for a short distance in order to avoid rapid expansion due to Coulomb forces between them [13], or they are accelerated via a DC field, then compressed via an RF field [14, 15]. There are also designs where acceleration and compression take place through the same RF field [16]. Furthermore, utilizing relativistic electron sources can also improve brightness and time resolution since they greatly reduce pulse broadening effects [17–19]. In all cases a short probe pulse is produced. Probe pulses capture molecular dynamics and produce a diffraction pattern. By varying the time delay between pump and probe pulses, the molecular dynamics at different time points can be captured and a'molecular movie' can be generated. As a result, the time resolution is mainly limited by the technological ability to produce ever-shorter laser and electron pulses [20], both for triggering a photo-induced dynamic rapidly and for imaging it. In other approaches, the probe pulse is dissected to increase time resolution. In the case of electrons, streak cameras [21] that spatially separate a long electron beam into smaller pieces have been developed, enabling higher time resolutions [22]. Another proposed method, known as optical gating [23], uses ultrashort laser pulses to dissect the electron beam and achieve a higher time resolution than was originally possible based on the length and speed of the beam.

For the sake of making this chapter as self-encompassing as possible, we will start with a review of special relativity (SR) and the concepts of time dilation and length contraction. This shall make for a smoother understanding of the core ideas of the proposed novel experiment.

2. Special relativity: a quick review

2.1 A brief history

Towards the end of the nineteenth century there was one major inconsistency plaguing the structure of theoretical physics. Newton's[1] equations described very well the mechanics of moving objects ranging from tiny objects on earth to the orbits of planets in space. Maxwell[2] had successfully completed the theoretical framework of electromagnetism in 1865, a monumental task that crowned the gradual understanding of electromagnetism throughout the eighteenth and nineteenth centuries through the work of scientists such as Coulomb[3], Ampère[4] and Faraday[5]. Despite the

---

[1] Sir Isaac Newton: English mathematician, physicist, engineer, philosopher, astronomer, theologian and author (1642–1726).

[2] James Clerk Maxwell: Scottish mathematician and physicist (1831–1879).

[3] Charles-Augustin de Coulomb: French engineer and physicist (1736–1806).

[4] André-Marie Ampère: French physicist and mathematician (1775–1836).

[5] Michael Faraday: English physicist (1791–1867).

enormous success of both Newton's and Maxwell's frameworks in describing mechanical objects and light, respectively, they were inconsistent with each other. Maxwell's equations predicted a constant speed of light and Newton's equations suggest that objects moving at different speeds should measure the speed of light to be different. On a more rigorous level, Newton's equations are invariant under Galilean[6] transformations while Maxwell's equations are invariant under Lorentz[7] transformations. So, each set of equations suggested a different symmetry present in nature. Moreover, in 1887, on the experimental side, Michelson[8] and Morley[9] tried to detect a difference in the speed of light for observers moving at different speeds but their results were decidedly negative. There were some attempts to resolve this consistency but it was Einstein,[10] who, in 1905, successfully presented the correct solution through a radical theory that would change our understanding of nature forever.

To resolve the inconsistency Einstein suggested that:

1. The laws of physics are invariant in all inertial frames of reference.

2. The speed of light in vacuum is constant in all frames of reference.

Simply put, he suggested that Newton's equations were fundamentally wrong and he replaced them with the equations of SR that were invariant under Lorentz transformations. The implications were dramatic: different observers experience different rates of time, lengths can shrink or elongate and many other peculiar effects take place as objects approach the speed of light. Newton's equations were only a very good approximation as long as objects moved slowly compared to the speed of light.

2.2 Mathematical framework

There are many ways to approach SR from a mathematical point of view. In this chapter we will present the mathematical framework of SR in a simple manner.

2.2.1 Galilean transformations

Limiting our consideration to one spatial dimension for simplicity, the most general way one can transform between two coordinate systems O and O', where O' is moving with speed v in the positive x-direction compared to O, is the following:

$$\begin{bmatrix} x' \\ t' \end{bmatrix} = \begin{bmatrix} a & b \\ e & f \end{bmatrix} \begin{bmatrix} x \\ t \end{bmatrix} \tag{1}$$

Assuming that space is homogeneous and noticing that the principle of relativity requires that O moves at speed -v compared to O', we are restricted to linear transformations. A deeper mathematical analysis of this is outside the scope of this chapter.

---

6 Galileo Galilei: Italian astronomer, physicist, engineer, philosopher, and mathematician (1564–1642).

7 Hendrik Antoon Lorentz: Dutch physicist (1853–1928).

8 Albert Abraham Michelson: German-born American physicist (1852–1931).

9 Edward Williams Morley: American physicist (1838–1923).

10 Albert Einstein: German physicist (1862–1943).

Taking all this into account the forward and backward transformations are calculated to be:

$$\begin{bmatrix} x' \\ t' \end{bmatrix} = \begin{bmatrix} a & -av \\ \dfrac{1-a^2}{av} & a \end{bmatrix} \begin{bmatrix} x \\ t \end{bmatrix} \tag{2}$$

$$\begin{bmatrix} x \\ t \end{bmatrix} = \begin{bmatrix} a & av \\ \dfrac{a^2-1}{av} & a \end{bmatrix} \begin{bmatrix} x' \\ t' \end{bmatrix} \tag{3}$$

The Galilean transformations coincide with our everyday intuition ($a = 1$). Velocities are additive, acceleration is invariant and time is the same for all observers.

$$x' = x - vt \tag{4}$$

$$t' = t \tag{5}$$

2.2.2 Lorentz transformations

Going back to the general form of the transformations, we have:

$$x' = ax - bt \tag{6}$$

$$x = ax' + bt' \tag{7}$$

By setting $x' = 0$, we can calculate the relative velocity of O' with respect to O

$$V' = b/a \equiv v \tag{8}$$

and similarly, by setting $x = 0$, we calculate

$$V = -b/a = -v \tag{9}$$

Assuming the speed of light is constant in all inertial frames of reference, we consider a light signal when the origins coincide ($t' = t$, $x' = x$). The propagation of the light signal in both frames is:

$$x = ct \tag{10}$$

$$x' = ct' \tag{11}$$

Substituting Eqs. (10) and (11) into Eqs. (6) and (7) yields:

$$ct' = (ac - b)t \tag{12}$$

$$ct = (ac + b)t' \tag{13}$$

Substituting Eq. (13) into Eq. (12) and using Eq. (8) yields:

$$a = \frac{1}{\sqrt{1 - v^2/c^2}} \equiv \gamma \tag{14}$$

$$b = av = \gamma v \tag{15}$$

Substituting Eqs. (14) and (15) into Eqs. (6) and (7) yields:

$$x' = \gamma(x - vt) \tag{16}$$

$$x = \gamma(x' + vt') \tag{17}$$

By substituting Eq. (16) into Eq. (17) the transformation of time is obtained:

$$t' = \gamma\left(t - \frac{vx}{c^2}\right) \tag{18}$$

$$t = \gamma\left(t' + \frac{vx'}{c^2}\right) \tag{19}$$

2.2.3 Time dilation

Consider an observer in the frame of reference O at the origin, so $x = 0$ with a clock that has a period $\Delta t = t_2 - t_1$. For the observer in O' that period is much longer, namely

$$\Delta t' = \gamma \Delta t. \tag{20}$$

Hence, the clock is ticking slower for the moving observer.

2.2.4 Length contraction

Similar to the peculiar effect of time dilation, a moving observer experiences a contraction in length along the direction of movement. Consider an observer in the frame of reference O that measures a moving ruler to be of length Δx. This measurement happens instantly at one point in time such that $t_2 = t_1$. This is an important detail as simultaneous times in one frame are not simultaneous in the other one. For the observer in O' that length transforms to

$$\Delta x' = \gamma \Delta x. \tag{21}$$

Now, by symmetry, i.e., a ruler at rest in O, that is measured to be of length L_o in O, is measured to be of length

$$L_o' = \frac{1}{\gamma} \Delta L_o \tag{22}$$

in O'. Alternatively, we could have considered an object at rest in O at the beginning but this would have added an extra step in the derivation as we would have considered the transformations both in space and time. Either way we arrive at the same result; moving objects experience length contraction along the direction of motion.

With that knowledge in mind, we are now ready to discuss the proposed experiment.

3. The proposed experiment

This section presents a method for studying molecular and atomic dynamics using time-resolved diffraction or spectroscopy studies with greater time resolution without relying on laser or electron beam technology advancements [24]. In this method, instead of shortening the probe pulse, the 'internal clock' of the sample (charged molecule or ion) is slowed down. This can be accomplished by accelerating the sample to relativistic speeds, which can be realized in particle accelerators, such as cyclotrons and synchrotrons. A sample, which is accelerated to speed v_s, undergoes a slowing down of its 'internal clock' by a factor of γ, where

$$\gamma = 1/\sqrt{1 - v_s^2/c^2} \tag{23}$$

relative to the lab frame irrespective of its velocity direction. As a result, the time resolution becomes a function of the sample's energy rather than being mainly reliant on pump and probe pulse durations. This can easily enable new time resolutions that have never been unlocked before.

3.1 Experimental considerations

To successfully implement any novel method, there are several barriers and challenges to overcome. We will introduce the experimental setup and discuss the experimental challenges and limitations, as well as the physics involved in the proposed setup.

3.1.1 Setup

We propose accelerating the samples in a cyclotron or synchrotron and studying them at a fixed energy E and, consequently, at a fixed speed v_s, which remains constant during data collection. **Figure 1** shows a schematic of the experimental setup. During the experiment, samples are accelerated into a chamber in which a pump pulse

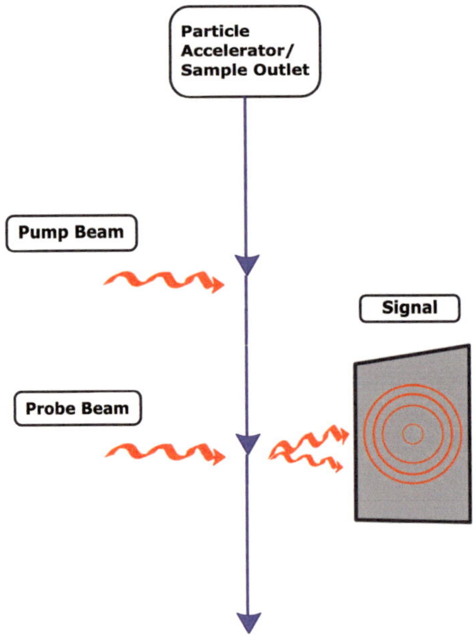

Figure 1.
Schematic of the proposed experimental setup: Samples are accelerated to a fixed energy. A pump pulse and a probe pulse are directed parallel to each other and perpendicular to the sample direction of motion. The delay between the pump pulse and probe pulse can be controlled by changing the distance between the two beams.

is directed parallel to a probe pulse and perpendicular to the direction of the sample's motion. The delay between the pump pulse and probe pulse can be controlled by changing the distance L between the two beams. The resulting time delay τ_d' according to the sample's clock is

$$\tau_d' = \frac{L}{v_s \gamma}. \tag{24}$$

As a result, the signal will reflect the changing dynamics according to the 'internal clock' of the sample.

3.1.2 Sample suitability

To begin with, this novel method can only be applied to electrically charged ions or molecules so that they can be accelerated to relativistic speeds, and to molecules that are in the gas phase in order to achieve the required energies. To run the experiment for a given time resolution, the higher the mass of the sample, the more energy is needed, so the best candidates for such studies are light, charged molecules, and ions.

The typical number density in gas phase UED [25–27] and spectroscopy [28] experiments is $\sim 10^{15}$ cm^{-3}. Proton bunches at the LHC contain 1.15×10^{11} protons with a proton number density of $\sim 10^{16}$ cm^{-3} [29, 30]. There have been many schemes to reduce bunch length, and in fact an order of magnitude shorter bunch length has been produced for the purpose of accelerating electrons with plasma wakefields of proton bunches [31].

It is also necessary for a sample to be stable when subjected to the accelerator conditions. H-anions with a binding energy of $\sim$ 0.75 eV are accelerated regularly to 520 MeV at TRIUMF [32]. Molecules with covalent bonds typically have binding energies of 1 eV or higher. The typical length scale of a covalent bond is 1 Å. Hence, to break a bond, a force of $\sim$1600 pN is needed. The forces exerted by typical electric fields (< 10 MV/m) and typical magnetic fields (< 10 T) are orders of magnitude less than 1600 pN. Furthermore, second ionization energies for atomic ions are typically significantly higher than 1 eV.

In addition to originally being designed to accelerate only protons and positively charged ions, the LHC ring has recently been used to accelerate partially stripped Pb^{+81} ions with one electron to an energy of 6.5 Q TeV [33] as part of the gamma factory proposal [34] to create a new type of high intensity light source. Although originally designed to accelerate protons or positively charged ions only, the LHC ring has recently accelerated partially stripped Pb^{+81} ions with one electron to an energy of 6.5 Q TeV [33], where Q is the ion charge number, as part of the gamma factory proposal [34] to create a new type of high intensity light source. Currently, engineering challenges with regard to collimation are being addressed [35].

3.2 Theoretical considerations

3.2.1 Energy considerations

At the moment, the Large Hadron Collider at CERN can accelerate protons to energies on the order of 7 TeV [36] and lead ions to 5 TeV collision energies [37], which is enough to boost the resolution of time measurements significantly. As an example, a hydronium molecule (H_3O^+, rest mass: 3.16×10^{-26} kg) accelerated to an energy of 1.8 TeV would experience a slowing down of time with a γ factor of 100. Since

$$E = \gamma mc^2, \tag{25}$$

the time resolution scales proportionally to energy, so an energy of 18 TeV would result in an astonishing $\gamma = 1000$. Additionally, time resolution is inversely proportional to mass, so hydrogen ions, for example, would experience an order of magnitude more gain in time resolution than hydronium molecules with the same energy.

The effect of relativistic time dilation on dynamical processes will still be extremely fascinating to observe, even if the particle accelerators cannot be commissioned to perform this experiment in the near future. With current laser technology it is possible to observe changes in differential detection, with and without a perturbation, as small as 10^{-4} to 10^{-8} [38, 39] using standard modulation techniques and photon detectors. There has also been major advances in laser based particle accelerators up to field gradients as high as 100 GeV/m [40–43] that will soon enable particle kinetic energies up to 10 GeV range or higher. As compared with particle accelerators, this level of relativistic energy would result in only modest time dilation. It would nonetheless constitute a direct measurement of time retardation, which would prove to be an important test case for the development of laser-based particle acceleration with the goal of controlling the time variable directly, asymptotically approaching'-stopping' time. A control of the time variable could open up new avenues, beyond simple imaging, to driving dynamics that are otherwise too rapid to control.

3.2.2 Pump and probe beam dynamics

As in standard ultrafast studies, pulsed pump and probe beams can be used. The time resolution is largely determined by the pulse duration of the pump and probe pulses. In conventional terms, the pulse duration refers to the time during which the full width at half maximum (FWHM) of the pulse crosses the sample. The pump pulse determines the trigger speed, and the probe pulse determines the imaging time resolution.

Besides velocity mismatch [44, 45], which takes place due to the difference in velocity between pump and probe pulses and their different incidence angles, other factors that affect the time resolution are the time of arrival jitter [25] for RF accelerated electron pulses. By the very nature of the experimental geometry, however, a lower resolution due to velocity mismatch or time of arrival jitter is avoided, as both pulses are parallel, so the delay time from time zero is solely determined by the speed of the sample between the pump and probe pulses. According to our proposed setup, the pulse crosses the sample in two directions, and we will thus consider the pulse duration during which the pulse crosses the sample or vice versa, in both directions. To explain the physics we denote the direction, parallel to the direction of propagation of the sample beam, y and denote the perpendicular direction x. For clarification, we will treat the problem from the lab frame of reference as well as from the sample frame of reference.

a. **Lab frame of reference.** The pulse duration in the x-direction τ_x is given by

$$\tau_x = \frac{l_x}{v_p}, \tag{26}$$

where l_x indicates the pulse length in the x-direction and v_p indicates the speed of the pump/probe pulse. The pulse duration in the y-direction τ_y is given by

$$\tau_y = \frac{l_y}{v_s}, \tag{27}$$

where l_y indicates the pulse length in the y-direction and v_s indicates the speed of the sample. V_s will always be close to the speed of light c in the proposed experiment.

b. **Sample frame of reference.** With relativistic speeds approaching the speed of light, the effect of length contraction along the direction of sample propagation y becomes significant. Hence, the length of the pulse in the y-direction is contracted to

$$l'_y = \frac{l_y}{\gamma}. \tag{28}$$

The pulse duration in the y-direction τ'_y is then given by

$$\tau'_y = \frac{l_y}{\gamma v_s}. \tag{29}$$

The pulse duration in the x-direction τ'_x is given by

$$\tau'_x = \frac{\gamma l_x}{v_p} \tag{30}$$

as v_p transforms to

$$v'_p = \frac{v_p}{\gamma} \tag{31}$$

in the sample frame of reference. This happens because of relativistic angle aberration. In the lab frame of reference a pulse that is emitted at 90° does not hit the sample perpendicularly in the sample frame of reference. The closer v_s is to c the smaller is the incidence angle between the beam and the sample's line of motion in the sample frame of reference. When discussing the observable signal relativistic angular aberration will be discussed in more detail.

However, as a quick check, if we assume that the probe beam consists of photons, then $v_p = c$. Utilizing Eq. (41), the longitudinal component of velocity in the sample frame of reference is given by $c\cos\theta'_i = c\frac{\cos\theta_i - \frac{v}{c}}{1-\cos\theta_i\frac{v}{c}}$, where θ_i is the angle in the lab frame of reference and v is the speed of the sample in the lab frame of reference. Plugging in $\theta_i = \frac{\pi}{2}$, we end up with $-v$ as the longitudinal component. The total speed of the photon probe pulse is then given by $\sqrt{\left(\frac{c}{\gamma}\right)^2 + (-v)^2} = \sqrt{c^2\left(1-\frac{v^2}{c^2}\right) + (-v)^2} = c$, which is expected as the speed of light is constant in all frames of reference.

Typical pulses are Gaussian temporally and spatially ($\tau_x \cong \tau_y$) and so the time resolution τ_{res} would be determined by the relativistically shortened duration τ'_y. It would thus be given by

$$\tau_{res} = \frac{1}{\gamma}\sqrt{\tau_{pump}^2 + \tau_{probe}^2}, \tag{32}$$

where τ_{pump} and τ_{probe} are the transit time durations of the sample beam through the pump and probe pulses in the lab frame, respectively. As an example for pump and probe beams with $l_y = 10\ \mu$m and $\gamma = 100$, the time resolution would be roughly 470 as. **Figure 2** shows a visual representation from both frames of reference along the relevant direction y.

The lab frame of reference is the same as the sample frame of reference in conventional pump-probe experiments. Signals (e.g., diffraction patterns) always reflect interaction time according to the clock rate of the sample, so our proposed setup exploits the involved relativistic effects that result from the differences between two frames of reference.

3.2.3 Doppler effect and frequency shifts

In order to properly conduct the experiment, it is imperative to understand how the frequency of a laser, x-ray or electron pulse is 'seen' by the sample in its own rest frame. Changing the frequency of a laser can cause it to be outside the absorption spectrum of the sample, preventing the intended interaction. The spatial resolution of scattering x-rays and electrons would decrease if they undergo significant redshifts, for example. Furthermore, there is the relativistic effect of time dilation in addition to the classical Doppler effect. Even if the source and receiver are not crossing paths, relativity dictates a frequency shift known as the transverse Doppler effect [46].

If we let θ be the angle between the sample wave vector and the wave vector of the pump/probe particles, as measured in the lab frame of reference, then the frequency

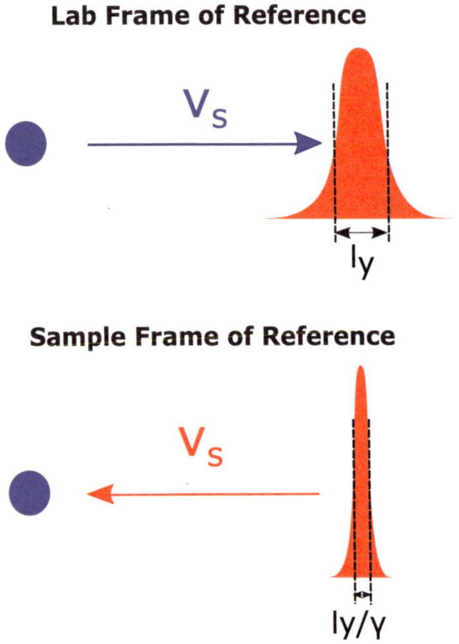

Figure 2.
Lab and sample frames of reference: Along the y-direction the length of the pulse is contracted in the sample frame of reference relative to the lab frame of reference.

that the sample 'sees', f_s, is given in terms of the frequency in the lab frame of reference, f_l, by

$$f_s = f_l\gamma(1 - \beta\cos\theta) \tag{33}$$

for photons (see derivation in Appendix A) [47]. For other particles, e.g., electrons, one needs to replace β with

$$\beta_e = \frac{v_s}{v_e}, \tag{34}$$

where v_e is the speed of the particles but $\gamma = 1/\sqrt{1 - v_s^2/c^2}$ remains the same. In our proposed setup, where the pump and probe beams are perpendicular to the direction of motion of the samples, we have

$$f_s = f_l\gamma. \tag{35}$$

Depending on the angle, there could either be a redshift or a blueshift. For one critical angle θ_c there is no frequency shift. For β (β_e) ≈ 1, as γ becomes larger, θ_c becomes smaller, meaning that the two wave vectors are more collinear. Although this would eliminate frequency shifts entirely, it would decrease the time resolution significantly.

3.2.4 The observable signal

The interaction between light and matter can take many different forms (e.g., absorption, scattering, etc.). In this section we present a general scheme for calculating the final observable signal in our proposed experiment.

The main steps are: (1) transforming the incident field from the lab frame of reference to the sample frame of reference by applying the Lorentz transformations; (2) calculating the resultant signal in the sample frame of reference; (3) transforming the signal from the sample frame of reference to the lab frame of reference by applying the Lorentz transformations one more time.

Without loss of generality, for an incident electric field E_i with angular frequency ω_i polarized in the z-direction and incident at angle θ_i (angle between photon wave vector $\vec{k}_i$ and sample velocity vector $\vec{v}_s$), the field would be Lorentz transformed to the sample frame of reference to E_i' in the following way: [48, 49].

$$\vec{E}_i = \hat{z}E_i \exp\left[ik_i(\cos\theta_i y + sin\,\theta_i x)\right]\exp\left(-i\omega_i t\right), \tag{36}$$

$$\vec{E}_i' = \hat{z}'E'_i \exp\left[ik_i'(\cos\theta_i' y' + sin\,\theta_i' x')\right]\exp\left(-i\omega_i' t'\right), \tag{37}$$

where

$$E_i' = \gamma(1 - \beta\cos\theta_i)E_i, \tag{38}$$

$$k_i' = \frac{\omega_i'}{c}, \tag{39}$$

$$\omega_i' = \gamma(1 - \beta\cos\theta_i)\omega_i, \tag{40}$$

$$\theta_i' = \cos^{-1}\left[\frac{\cos\theta_i - \beta}{1 - \cos\theta_i(\beta)}\right]. \tag{41}$$

For particles other than photons moving with speed v_e the angular frequency and angle transform in the following way:

$$\omega_i' = \gamma(1 - \beta_e \cos\theta_i)\omega_i, \tag{42}$$

$$\theta_i' = \tan^{-1}\left[\frac{\sin\theta_i}{\gamma(\cos\theta_i - \beta_e)}\right]. \tag{43}$$

However, as before $\gamma = 1/\sqrt{1-\beta^2}$ with $\beta = \frac{v_s}{c}$ remains the same.

Due to the nature of the angular transformations (see derivation in Appendix B), we expect scattering and diffraction angles to be wider than for the static case, and hence, we recommend detectors that cover as much of the 4π sr solid angle as possible.

Momentum transfer due to light radiation pressure would have a more negligible effect on changing the ion beam path than in conventional ultrafast gas electron diffraction. Forces perpendicular to the ion beam path would cause acceleration according to $a_{transverse} = \frac{F_{transverse}}{\gamma m}$. Forces along the ion beam path would cause acceleration according to $a_{longitudinal} = \frac{F_{longitudinal}}{\gamma^3 m}$. Hence, acceleration due to transverse forces is reduced by a factor of $\frac{1}{\gamma}$ and acceleration due to longitudinal forces is reduced by a factor of $\frac{1}{\gamma^3}$ as compared to conventional non-relativistic ultrafast gas electron diffraction. This is due to the well-known concept of transverse and longitudinal masses in special relativity.

4. Conclusions

In this chapter, we proposed a novel method for time-resolved studies that relies on taking advantage of relativistic effects rather than on advances in laser technology or electron beam compression. It was shown that by using currently available technology, this method could improve time resolution by 2 or 3 orders of magnitude. This has the potential of opening up a whole new domain of ultrafast dynamics that was previously unattainable. With ever more powerful accelerators being proposed, there will be great potential to achieve truly remarkable time resolutions that far exceed the status quo. We hope this would lead to new avenues of collaboration between the particle physics community and the ultrafast science community, which will maximize the research potential of particle accelerator facilities worldwide.

Acknowledgements

Hazem Daoud would like to thank Prof. R.J. Dwayne Miller for his support, and Prof. Pierre Savaria and Prof. David Bailey for fruitful discussions.

Appendix A. Relativistic frequency shifts

We will derive the relativistic frequency shift formula in two steps. First we will consider a classical example to derive the classical Doppler shift then we will consider the relativistic effects involved in the second step.

A.1 Classical Doppler shift

Consider a resting source emitting a photon with frequency f_0 and period T_0 collinear with an observer moving with speed u in the same direction as the emitted photon. We denote the frequency observed as f_S. The observed wavelength λ_S is given by

$$\lambda_S = (c - u)T_0. \tag{44}$$

Hence,

$$\frac{c}{f_S} = \frac{c - u}{f_0}, \tag{45}$$

and so,

$$f_0 = (1 - \beta)f_S, \tag{46}$$

where $\beta = \frac{u}{c}$.

A.2 Relativistic Doppler shift

a. The time interval between consecutive crests in the source frame of reference is T_0.

b. Due to relativistic time dilation, the time interval between consecutive crests in the observer frame of reference is γT_0.

c. The time interval between the reception of consecutive crests in the source frame of reference is $T_0(1 - \beta \cos\theta)$, where θ is the angle between the observer's wave vector and the wave vector of the photon.

d. The time interval between the reception of consecutive crests in the observer frame of reference is $\gamma T_0(1 - \beta \cos\theta)$. A calculation analogous to the classical Doppler shift yields $f_0 = \gamma(1 - \beta \cos\theta)f_S$

Appendix B. Relativistic angular aberration

Consider a probe beam emitted from a source at an angle θ and speed u with respect to the wave vector of an observer moving with speed v in the frame of reference of the source. Considering a 2-d plane containing the wave vectors of the probe beam and the observer, we can define the components as follows:

$$u_x = u\cos\theta \tag{47}$$

$$u_y = u\sin\theta \tag{48}$$

In the observer frame of reference (indicated with a prime), using the relativistic velocity transformations, yields:

$$u'_x = \frac{u\cos\theta - v}{1 - \frac{v}{c^2}u\cos\theta} \tag{49}$$

$$u'_y = \frac{\sqrt{1 - \frac{v^2}{c^2}}u\sin\theta}{1 - \frac{v}{c^2}u\cos\theta} \tag{50}$$

Hence,

$$\tan\theta' = \frac{u'_x}{u'_y} = \frac{\sqrt{1 - \frac{v^2}{c^2}}u\sin\theta}{u\cos\theta - v}. \tag{51}$$

The probe beam speed in the observer frame of reference is

$$u' = \sqrt{u'^2_x + u'^2_y} = \frac{\left(u^2 - \frac{u^2v^2\sin\theta^2}{c^2} - 2u\cos\theta v + v^2\right)^{1/2}}{1 - \frac{v}{c^2}u\cos\theta}.\widehat{\omega} \tag{52}$$

So,

$$\cos\theta' = \frac{u'_x}{u'} = \frac{u\cos\theta - v}{\left(u^2 - \frac{u^2v^2\sin\theta^2}{c^2} - 2u\cos\theta v + v^2\right)^{1/2}} \tag{53}$$

If the probe beam consists of photons, then $u = c$ and this simplifies further to

$$\cos\theta' = \frac{\cos\theta - \frac{v}{c}}{1 - \frac{v}{c}\cos\theta}. \tag{54}$$

Author details

Hazem Daoud
University of Toronto, Toronto, Canada

*Address all correspondence to: hazem.daoud@mail.utoronto.ca

References

[1] Miller RJD. Femtosecond crystallography with ultrabright electrons and x-rays: Capturing chemistry in action. Science. 2014; **343**(6175):1108-1116

[2] Dwayne Miller RJ. Mapping atomic motions with ultrabright electrons: Realization of the chemists' Gedanken experiment. In: 13th International Conference on Fiber Optics and Photonics. Washington, D.C., United States: Optical Society of America; 2016. p. P1A.25

[3] Sciaini G, Miller RD. Femtosecond electron diffraction: heralding the era of atomically resolved dynamics. Reports on Progress in Physics. 2011;**74**(9): 096101

[4] Polanyi JC, Zewail AH. Direct observation of the transition state. Accounts of Chemical Research. 1995; **28**(3):119-132

[5] Corkum PÁ, Krausz F. Attosecond science. Nature Physics. 2007;**3**(6): 381-387

[6] Barends TRM, Foucar L, Ardevol A, Nass K, Aquila A, Botha S, et al. Direct observation of ultrafast collective motions in co myoglobin upon ligand dissociation. Science. 2015;**350**(6259): 445-450

[7] Altarelli M, editor. XFEL, the European X-ray Free-Electron Laser: Technical Design Report. Hamburg: DESY XFEL Project Group; 2006. ISBN 9783935702171. OCLC: 254657183

[8] Siwick BJ, Dwyer JR, Jordan RE, Miller RD. Ultrafast electron optics: Propagation dynamics of femtosecond electron packets. Journal of Applied Physics. 2002;**92**(3):1643-1648

[9] Ruan CY, Lobastov VA, Vigliotti F, Chen S, Zewail AH. Ultrafast electron crystallography of interfacial water. Science. 2004;**304**(5667):80-84

[10] Rosker MJ, Dantus M, Zewail AH. Femtosecond clocking of the chemical bond. Science. 1988;**241**(4870): 1200-1202

[11] Mokhtari A, Cong P, Herek JL, Zewail AH. Direct femtosecond mapping of trajectories in a chemical reaction. Nature. 1990;**348**(6298):225-227

[12] Emma P, Akre R, Arthur J, Bionta R, Bostedt C, Bozek J, et al. First lasing and operation of an angstrom-wavelength free-electron laser. Nature Photonics. 2010;**4**(9):641-647

[13] Siwick BJ, Dwyer JR, Jordan RE, Miller RD. An atomic-level view of melting using femtosecond electron diffraction. Science. 2003;**302**(5649): 1382-1385

[14] Veisz L, Kurkin G, Chernov K, Tarnetsky V, Apolonski A, Krausz F, et al. Hybrid dc–ac electron gun for fs-electron pulse generation. New Journal of Physics. 2007;**9**(12):451-451

[15] Van Oudheusden T, De Jong EF, Van der Geer SB, Root WO, Luiten OJ, Siwick BJ. Electron source concept for single-shot sub-100 fs electron diffraction in the 100 kev range. Journal of Applied Physics. 2007;**102**(9): 093501

[16] Daoud H, Floettmann K, Dwayne Miller RJ. Compression of high-density 0.16 pc electron bunches through high field gradients for ultrafast single shot electron diffraction: The compact rf gun. Structural Dynamics. 2017;**4**(4): 044016

[17] Hastings JB, Rudakov FM, Dowell DH, Schmerge JF, Cardoza JD, Castro JM, et al. Ultrafast time-resolved electron diffraction with megavolt electron beams. Applied Physics Letters. 2006;**89**(18):184109

[18] Murooka Y, Naruse N, Sakakihara S, Ishimaru M, Yang J, Tanimura K. Transmission-electron diffraction by mev electron pulses. Applied Physics Letters. 2011;**98**(25):251903

[19] Musumeci P, Moody JT, Scoby CM, Gutierrez MS, Westfall M, Li RK. Capturing ultrafast structural evolutions with a single pulse of mev electrons: Radio frequency streak camera based electron diffraction. Journal of Applied Physics. 2010;**108**(11):114513

[20] Baum P, Zewail AH. Breaking resolution limits in ultrafast electron diffraction and microscopy. Proceedings of the National Academy of Sciences. 2006;**103**(44):16105-16110

[21] Kassier GH, Haupt K, Erasmus N, Rohwer EG, Von Bergmann HM, Schwoerer H, et al. A compact streak camera for 150 fs time resolved measurement of bright pulses in ultrafast electron diffraction. Review of Scientific Instruments. 2010;**81**(10):105103

[22] Badali DS, Dwayne Miller RJ. Robust reconstruction of time-resolved diffraction from ultrafast streak cameras. Structural Dynamics. 2017; **4**(5):054302

[23] Hassan MT, Baskin JS, Liao B, Zewail AH. High-temporal-resolution electron microscopy for imaging ultrafast electron dynamics. Nature Photonics. 2017;**11**(7):425-430. ISSN 1749-4885, 1749-4893

[24] Daoud H, Miller RD. Utilizing relativistic time dilation for time-resolved studies. The Journal of Chemical Physics. 2021;**154**(11):111107

[25] Centurion M. Ultrafast imaging of isolated molecules with electron diffraction. Journal of Physics B: Atomic, Molecular and Optical Physics. 2016;**49** (6):062002

[26] Yang J, Centurion M. Gas-phase electron diffraction from laser-aligned molecules. Structural Chemistry. 2015; **26**(5–6):1513-1520

[27] Shen X, Nunes JPF, Yang J, Jobe RK, Li RK, Lin MF, et al. Femtosecond gas-phase mega-electron-volt ultrafast electron diffraction. Structural Dynamics. 2019;**6**(5):054305

[28] Brünken S, Lipparini F, Stoffels A, Jusko P, Redlich B, Gauss J, et al. Gas-phase vibrational spectroscopy of the hydrocarbon cations l-c3h+, hc3h+, and c-c3h2+: Structures, isomers, and the influence of ne-tagging. The Journal of Physical Chemistry A. 019;**123**(37): 8053-8062

[29] Zimmermann F, Assmann R, Giovannozzi M, Papaphilippou Y, Caldwell A, Xia G. Generation of short proton bunches in the CERN accelerator complex. In: Particle Accelerator Conference (PAC 09), page FR5RFP004. Vancouver, BC, Canada: INSPIREHEP; 2010

[30] Steerenberg R. LHC Report: Full House for the LHC. 2017

[31] Adli E, Ahuja A, Apsimon O, Apsimon R, Bachmann A-M, Barrientos D, et al. Acceleration of electrons in the plasma Wakefield of a proton bunch. Nature. 2018;**561**(7723): 363-367

[32] TRIUMF. Main Cyclotron and Beam Lines [Accessed: January 27, 2021]

[33] Schaumann M, Alemany-Fernandez R, Bartosik H, Bohl T, Bruce R, Hemelsoet GH, et al. First partially stripped ions in the lhc (208pb81+). Journal of Physics: Conference Series. 2019;**1350**:012071

[34] Krasny MW. The Gamma factory proposal for CERN. PoS, EPS-HEP. 2018; **2017**:532

[35] Gorzawski A, Abramov A, Bruce R, Fuster-Martinez N, Krasny M, Molson J, et al. Collimation of partially stripped ions in the cern large hadron collider. Physical Review Accelerators and Beams. 2020;**23**(10): 101002

[36] Fartoukh S, Karastathis N, Ponce L, Camillocci MS, Garcia RT. About flat telescopic optics for the future operation of the LHC. CERN Reports, 2020. 2018; **8**:101002

[37] Foka P, Janik MA. An overview of experimental results from ultra-relativistic heavy-ion collisions at the CERN LHC: Hard probes. Reviews in Physics. 2016;**1**:172-194

[38] Koke S, Grebing C, Frei H, Anderson A, Assion A, Steinmeyer G. Direct frequency comb synthesis with arbitrary offset and shot-noise-limited phase noise. Nature Photonics. 2010; **4**(7):462-465

[39] Quinlan F, Fortier TM, Jiang H, Hati A, Nelson C, Fu Y, et al. Exploiting shot noise correlations in the photodetection of ultrashort optical pulse trains. Nature Photonics. 2013; **7**(4):290-293

[40] Esarey E, Schroeder CB, Leemans WP. Physics of laser-driven plasma-based electron accelerators. Reviews of Modern Physics. 2009;**81**: 1229-1285

[41] Leemans WP, Gonsalves AJ, Mao HS, Nakamura K, Benedetti C, Schroeder CB, et al. Multi-GeV electron beams from capillary-discharge- guided subpetawatt laser pulses in the self-trapping regime. Physical Review Letters. 2014;**113**(24): 245002

[42] Joshi C, Corde S, Mori WB. Perspectives on the generation of electron beams from plasma-based accelerators and their near and long term applications. Physics of Plasmas. 2020; **27**(7):070602

[43] Blumenfeld I, Clayton CE, Decker F-J, Hogan MJ, Huang C, Ischebeck R, et al. Energy doubling of 42 gev electrons in a metre-scale plasma wakefield accelerator. *Nature*. 2007;**445**(7129): 741-744

[44] Williamson JC, Zewail AH. Ultrafast electron diffraction. Velocity mismatch and temporal resolution in crossed-beam experiments. Chemical Physics Letters. 1993;**209**(1-2):10-16

[45] Goodno GD, Astinov V, Miller RD. Diffractive optics-based heterodyne-detected grating spectroscopy: application to ultrafast protein dynamics. The Journal of Physical Chemistry B. 1999;**103**(4):603-607

[46] Mandelberg HI, Witten L. Experimental verification of the relativistic Doppler effect. JOSA. 1962; **52**(5):529-535

[47] Klein R, Grieser R, Hoog L, Huber G, Klaft I, Merz P, et al. Measurement of the transverse doppler shift using a stored relativistic7li ion beam. Zeitschrift für Physik A Hadrons and Nuclei. 1992; **342**:455-461

[48] Lee SW, Mittra R. Scattering of electromagnetic waves by a moving

cylinder in free space. Canadian Journal of Physics. 1967;**45**(9):2999-3007

[49] Papas CH. Theory of Electromagnetic Wave Propagation. New York, NY, USA: McGraw-Hill Book Co.; 1965